Principles and Practices
of Aquatic Law

Principles and Practices of Aquatic Law

By
John R. Fletemeyer
with Contributors

Edited by
Ivonne Schmid

CRC Press is an imprint of the
Taylor & Francis Group, an **informa** business

CRC Press
Taylor & Francis Group
6000 Broken Sound Parkway NW, Suite 300
Boca Raton, FL 33487-2742

© 2018 by Taylor & Francis Group, LLC
CRC Press is an imprint of Taylor & Francis Group, an Informa business

No claim to original U.S. Government works

Printed on acid-free paper

International Standard Book Number-13: 978-1-138-06652-6 (Hardback)

This book contains information obtained from authentic and highly regarded sources. Reasonable efforts have been made to publish reliable data and information, but the author and publisher cannot assume responsibility for the validity of all materials or the consequences of their use. The authors and publishers have attempted to trace the copyright holders of all material reproduced in this publication and apologize to copyright holders if permission to publish in this form has not been obtained. If any copyright material has not been acknowledged please write and let us know so we may rectify in any future reprint.

Except as permitted under U.S. Copyright Law, no part of this book may be reprinted, reproduced, transmitted, or utilized in any form by any electronic, mechanical, or other means, now known or hereafter invented, including photocopying, microfilming, and recording, or in any information storage or retrieval system, without written permission from the publishers.

For permission to photocopy or use material electronically from this work, please access www.copyright.com (http://www.copyright.com/) or contact the Copyright Clearance Center, Inc. (CCC), 222 Rosewood Drive, Danvers, MA 01923, 978-750-8400. CCC is a not-for-profit organization that provides licenses and registration for a variety of users. For organizations that have been granted a photocopy license by the CCC, a separate system of payment has been arranged.

Trademark Notice: Product or corporate names may be trademarks or registered trademarks, and are used only for identification and explanation without intent to infringe.

Visit the Taylor & Francis Web site at
http://www.taylorandfrancis.com

and the CRC Press Web site at
http://www.crcpress.com

Contents

Introduction vii
Principal Author xix
Contributors xxi

Chapter 1   An Overview of Aquatic Law 1
Michael Flynn

Chapter 2   Legal Concepts, Drowning, and Lifeguard Effectiveness 25
John R. Fletemeyer

Chapter 3   Principles of Aquatic Risk Management from a Legal Perspective ... 49
John R. Fletemeyer

Chapter 4   Evidence Collection and the Daubert Standard 63
Lori St John

Chapter 5   Swimming Pool and Spa Safety 73
John R. Fletemeyer

Chapter 6   Headfirst Recreational Aquatic Accidents 99
Tim O'Brien

Chapter 7   Ocean Beaches, Lakes, Quarries, Springs, and Canals 137
John R. Fletemeyer

Chapter 8   Commercial Whitewater Rafting 169
Julie Munger and Abigail Polsby

Chapter 9   *Breaux v. City of Miami Beach*, 899 So.2d 1059 (Fla. 2005) 197
Howard Pomerantz

Chapter 10 Past Its Prime: The 1920 Death on the High Seas Act 209
Michael D. Eriksen

Chapter 11 Justice for Deaths Caused by Free-Living Amebae
in Recreational Waters 223

Roger W. Strassburg, Jr.

Chapter 12 The Psychological Evaluation: Trauma Caused by a Serious
Aquatic Injury or Fatal Drowning 237

Diana P. Sullivan

Chapter 13 Drowning Forensics, Drowning and Accident Reenactments,
and Case Research 249

John R. Fletemeyer

Chapter 14 Boating and SCUBA Accidents, Low-Head Dams, and Beach
Renourishment 261

John R. Fletemeyer

Chapter 15 Hydrodynamic Analysis of Drowning Events 281

Brian K. Haus

Index 291

Introduction

PRACTICES AND PRINCIPLES OF AQUATIC LAW

More than a century ago, medicine and law began branching into several specialty fields and show no evidence of stopping. Law specialties include criminal law, patent law, constitutional law, health care law, and business law. The authors of this book are convinced that aquatic law deserves its own specialty recognition, and we are hopeful that the abundance of information presented in this book strongly supports this belief, not only by law practitioners but also by other professionals, including those working in aquatics, those working in risk management, and insurance underwriters.

In addition, we hope that this book will quickly become a major reference for practicing attorneys and a source to assist developing theories of negligence. We hope it will be similarly helpful to attorneys faced with the often formidable task of formulating defense strategies. If this book accomplishes this, then I will consider it a success even after factoring in more than five years of hard and often challenging work—not only by me, but by my colleagues who contributed unselfishly to this book.

It cannot be denied that fatal drownings, non-fatal drownings, and serious aquatic accidents are responsible for exacting an enormous economic and psychological toll on American society (see Figure I.1). Until relatively recently, only the economic toll was considered. Through anecdotal evidence, the psychological toll has gained increased consideration, especially among attorneys with high-value cases alleging negligence.

Take, for example, a young man who suffered a severe spinal injury after diving from a swimming pier that lacked shallow water warning signs and his subsequent attempts to commit suicide; and the teenaged lifeguard working at a public pool who failed to see a young boy drown while on his watch. This young man dropped out of college and experienced severe states of depression following this fatal and allegedly preventable drowning, with the fault placed squarely on this teen's shoulders. There is evidence that many young men and women experience posttraumatic stress disorder (PTSD) following a serious aquatic accident or fatal drowning. In the past, this debilitating condition was only reserved to men and women in the military. Not so anymore!

Arguably, of all the categories of accidental injury, drowning is the most preventable. As the most preventable accidental injury, a fatal or near-drowning is often the subject of a lawsuit alleging negligence or, in some especially egregious cases, gross negligence. Developing a reasonable theory of negligence and defending against a suit alleging negligence is part art and part science. This represents one of the themes of this book and this has been an observation that Professor of Law Michael Flynn, author of Chapter 1, has made repeatedly.

FIGURE I.1 Fatal drownings are responsible for an enormous social, psychological, and economic cost to society. Consider the multiple-drowning event on American Beach near Jacksonville, Florida, when during a family reunion, five family members drowned and ten others nearly drowned.

In recent years, juries have awarded large sums to wrongful death cases and to victims suffering permanent disabilities. In some cases, when gross negligence has been proven, punitive damages have been awarded, further increasing monetary amounts. Even in cases without serious injuries, victims often receive large settlements rather than risking an unfavorable jury award. As every seasoned attorney knows, jury decisions are never predictable. Consequently, most aquatic injury and drowning cases are settled out of court but not before a long and arduous period of discovery and responding to interrogatories; and possibly writing or arguing against summary judgments and deposition testimony that often include opinions from a host of aquatic, economic, psychological, and medical experts.

Suits involving aquatic accidents are especially challenging because of the dynamic nature of most aquatic environments that account for about two-thirds of our planet's surface. In addition, and making the jobs of aquatic professionals difficult, there is a lack of uniform and often confusing codes related to how the industry and its practitioners must legally respond to aquatic hazards and safety. For example, in regards to swimming pools, there are no uniform codes. Each state has assumed the responsibility of adopting and enforcing its own regulations. Consequently, trying to reach a consensus about developing uniform standards, especially those related to warnings and to open water environments, is often problematic, and at the time of writing this book, unfortunately this has not been tackled.

In addition, aquatic suits are difficult because juries are usually composed of people without scientific backgrounds. Opinions from experts are usually founded in statistical and scientific information, thanks to the necessity of Daubert standards. Attorney Lori St John in Chapter 4 discusses the significance of this.

Introduction

Research conducted by the authors has identified several contributing factors associated with drowning (fatal and non-fatal). Sometimes these are the subject of negligence theories relating to the proximate cause, and in some cases, proximate causes. The most important of these are discussed in detail in the following chapters. They include the following:

- Lack of responsible adult supervision when a toddler or a child is in or near the water.
- Failure to understand and recognize behavior that accompanies an active drowning event, especially among lifeguards who have a duty to understand these and to be able to respond without delay.
- Failure to comply with model safety codes and standards.
- Failure to properly maintain and supervise swimming pools, including private, semi-public, and public pools.
- Failure to identify and respond to hazards by providing effective warnings based on ANSI standards.
- Failure by the public to comply with warnings and to use common and reasonable sense when bathing or when engaged in an aquatic activity.
- Failure of aquatic professionals to enforce rules related to conduct and safety.
- Alcohol abuse and drug intoxication. It is estimated that about 50 percent of aquatic accidents are alcohol-related.
- Bathing alone, especially in hot tubs and spas.
- Breath-holding games and competitions resulting in a condition called "shallow water blackout."
- Risk-taking behaviour, primarily among teenagers and young adult males.
- Trauma, with head and neck injury often being the most serious; failure to follow concussion injury protocol when a head, neck, or spinal injury is suspected.
- Syncope secondary to a medical condition (MI, seizures, diabetes, cerebrovascular accidents, arrhythmias, etc.).
- Diving into shallow water without the benefit of appropriate warnings.
- Failure of lifeguards, white water guides, management, and so on to have proper training and required current certification(s). This includes recurrent training to practice and rehearsing of first aid and rescue skills.
- Failure of lifeguards to identify a drowning and to make a timely response within industry-accepted standards.
- Failure by an aquatic facility (including management and staff) to conduct a comprehensive risk management assessment and to have an emergency action plan.

How Many People Drown?

Despite the number of fatal drownings reported by the Centers for Disease Control and Prevention (CDC), and contrary to what most people believe, the number is not precisely known. The CDC estimates that it is about 4,000, or a little less. However, this number may be seriously underestimated for several obvious and not-so-obvious reasons. Avramidis (2007) discovered that many fatal drownings lay "hidden" because of the inconsistent way that they are coded by medical examiners. A fatal drowning that receives a prioritized cardiac arrest code will not likely be recorded, nor will it be represented as a drowning statistic.

During a story featured on *60 Minutes* about Mexican and South American migrants drowning in the All-American canal and Rio Grande River, a study reported that over a ten-year period, at least 500 migrants fatally drowned in these two dangerous water environments (Fletemeyer, 2010). None of these victims were recorded as drowning statistics. There is evidence that hundreds of Haitians and Cubans have drowned in Florida waters while rafting and sailing across the Florida Straits in derelict and seriously overloaded boats. These tragic deaths are also seldom recorded as drowning statistics.

In addition, anytime someone suffers a submergence event and dies 24 hours or more later, it is classified as a secondary drowning and not as a fatal drowning (Fletemeyer, 1999). During a boating accident, if someone falls off a boat and drowns, the United States Coast Guard often reports this as a boating fatality and not as a fatal drowning. According to Coast Guard statistics, about 400 people drown from boats in America every year.

The number of non-fatal drownings (once called "near drownings") is even more problematic to estimate. The National Association of Rescue Divers reports that non-fatal drownings occur 500 to 600 times more frequently than fatal drownings. The CDC states that for every fatal drowning, there are four to five non-fatal drownings that require emergency room intervention and sometimes long-term hospitalization. This was recently confirmed by Dr. Anthony Ard during his paper given at the 2014 Aquatic Law Symposium in Ft. Lauderdale.

Within this backdrop, *Principles of Aquatic Law* was proposed and then written by several highly qualified contributors, all leading and recognized experts in their respective fields. The authors of this book were the first to recognize, and not without some regret, that this book will and must always be regarded as a work in progress because of the inherently dynamic nature of the aquatics industry and because of the ephemeral nature of aquatic hazards that are responsible for impacting public safety.

Aquatic hazards are constantly changing. They appear, disappear, and then reappear, sometimes in unusual and unexpected ways. Several chapters in this book are devoted to specific hazards, some only recently appearing in aquatic environments. Take, for example, the brain-eating ameba that has killed more than 100 bathers (refer to Chapter 11) and a condition called "shallow water blackout," which is implicated in more and more fatal drownings.

The need for this book was first recognized in 1992 during the first aquatic law conference and has been reiterated several times during subsequent aquatic law conferences and sea symposiums hosted by the International Swimming Hall of Fame, the Sea Grant Foundation, the American Red Cross, the National Oceanographic and Atmospheric Administration (NOAA), several universities, and the Florida Beach Patrol Chiefs Association. Recently another organization has been added to this list: The National Drowning Prevention Alliance (NDPA).

Introduction

During the first aquatic law symposium, there was an immediate recognition that accurate and objective information about drowning, aquatic injuries, and aquatic hazard warnings is not readily available but nevertheless needed not only by practitioners representing the legal profession but also by insurance underwriters, risk managers, and individuals involved in aquatics who might be involved in defending lawsuits alleging negligence. This need also harkens back to a fact mentioned earlier: That the majority of aquatic injuries are considered preventable and are, therefore, candidates for legal action.

To reiterate, this is a book in progress. The aquatics industry is becoming increasingly more proactive and is constantly evolving with new paradigms, especially involving lifeguard practices, emergency first aid, CPR procedures, and public education initiatives. Take, for example, the Virginia Grahame Baker Act. The adoption of the Virginia Grahame Baker Act makes it mandatory for all pools to have anti-entrapment drains. This is perhaps the best example of a new paradigm and how the aquatic industry is responding to new challenges and becoming more proactive in regard to aquatic safety.

Recognizing the importance of public education marks yet another important paradigm shift credited with saving countless lives. Programs focusing on providing swimming lessons to every American, especially to minority children, is yet another example of an important change. So is the trend for lifeguards to focus more on taking preventative lifesaving measures, as opposed to sitting in their towers and waiting for drownings to happen.

New case law is being written by the courts and, consequently, is being used to develop new and sometimes creative theories of negligence and malpractice. Sometimes new case law is being used to help write summary judgment doctrines. In addition, and sometimes unexpectedly, old case law is being revisited, reexamined, and interpreted in different and sometimes creative ways. In Chapter 2, and during the last three aquatic law symposiums, Professor Michael Flynn has painstakingly examined literally hundreds of cases and is able to present an impressive overview of the impact on drowning from a legal perspective. Attorney Howard Pomerantz devotes an entire chapter (Chapter 9) to discussion of how new case law resulting from a drowning on Miami Beach has influenced many new cases and impacted legal decisions by the courts.

It goes without saying that every attorney's responsibility is to provide their client with the best possible representation. Consequently, it is incumbent that both defense and plaintiff attorneys have, at the very least, a minimum understanding of aquatics and the plethora of hazards that can occur in aquatic environments, including swimming pools, water theme parks, lakes, rivers, canals, ponds, and fresh and salt water beaches. In addition, it is also incumbent for attorneys to know how the aquatics industry is expected and sometimes legally required to respond to these hazards.

Little, if any, convincing was needed to persuade the contributing writers that this should be the main objective of this book. The second objective took a little more convincing. It was to make this book the curriculum source for future aquatic law symposiums that have been regularly conducted by the Aquatic Law and Safety Institute since 1992. Hopefully, and soon, there will an online course for attorneys and paralegals that will feature this book as the curriculum. Once accomplished, this book and the aquatic law program will be available to every practicing attorney in

America. The authors of this book also hope that law students will benefit and that this book will appear in courses at law schools.

There is yet another objective of this book. It is to identify the many recreational activities that are conducted in aquatic environments and their associated hazards with these activities. After identifying these hazards, several following chapters provide discussion about the legal and ethical responsibilities that the aquatic industry has to respond to them. Sometimes this responsibility is clear and well defined with established codes and regulations; sometimes it is established by common sense and "reasonable" practice. This third objective has been accomplished in several chapters. For example, Olympic Coach Tim O'Brien (Chapter 6) discusses hazards associated with swimming pool diving, Dr. Brian Haus (Chapter 15) discusses hazards associated with rip currents, and Julie Munger and Abigail Polsby (Chapter 8) discuss the hazards associated with whitewater rafting.

There are far too many aquatic activities to list in this introduction, but some of the major pursuits include recreational and competition swimming, springboard and platform diving, SCUBA diving, sail and power boating (including using personal water crafts), whitewater rafting, surfing, and kayaking. Each of these activities has certain dangers along with inherent risks. It is, therefore, incumbent that attorneys become aware of how the aquatic industry has a legal and sometimes "reasonable" duty to respond to them. It is important to note that even as this book was being written, several new activities and safety hazards were added to the list. With America's improving economy and with more leisure time spent in the water, the number of fatal drownings and serious accidents will surely increase and concurrently, so will suits alleging negligence.

The range of aquatic hazards that injure bathers is truly remarkable. Some may be even a little surprising! Many life-threatening hazards have been known ever since bathing was culturally popularized before the turn of the nineteenth century (Figure I.2).

Rip currents are the most deadly of all the beach hazards, accounting for at least 100 fatal drownings on American beaches every year (Lushine et al., 1999). A section of Chapter 2 by Fletemeyer and the entirety of Chapter 15 by Haus are devoted to rip currents. The threat of rip currents to public safety was not recognized by the American public a hundred years ago. However, when the famous painter Winslow Homer painted a drowned victim being pulled from the rough surf in 1898 that he incorrectly titled, *Riptide*, the public quickly became aware of this deadly hazard. Thanks to new research and better education programs, most now call these deadly currents by their proper name—"rip currents," not "riptides." Hopefully, members of the bathing public are learning how to identity and avoid rip currents that frequently occur on nearly all ocean beaches and even on a couple of fresh water beaches. On beaches, whether public or private, if warnings are not posted about rip currents, then the stakeholders may be found negligent (see Figure I.3).

Introduction

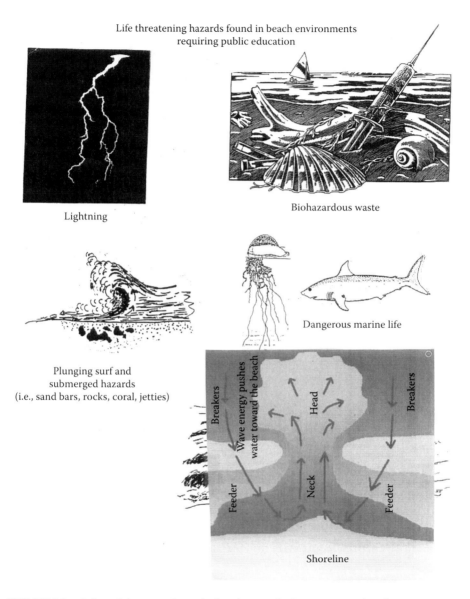

FIGURE I.2 A few of the many hazards that threaten bathers on ocean beaches.

FIGURE I.3 Warning signs and hazard flags are commonly used on most beaches to warn the public that rip currents may be present.

Cardio Pulmonary Resuscitation (CPR)

The importance of CPR was recognized in 1857 when Dr. Marshall Hall, an English physiologist, published a paper titled "Prone and Postural Respiration in Drowning." The following year, in 1858, Dr. Henry Sylvester devised a method of artificial respiration (Freas, 1999). However, the importance of CPR was not fully appreciated by the public until the American Red Cross began studying and demonstrating the efficacy of CPR. Once the effectiveness of CPR was established, the American Red Cross started a national campaign aimed at training every American in basic CPR procedures.

Since this campaign began, CPR protocols and practices have changed dramatically and frequently. This marks yet another paradigm shift, especially among ocean, open-water lifeguards. Once lifeguards were required to have only the basic first aid and CPR training. However, this has changed and now lifeguards are required to have advanced training, often at the EMT level, including specialty training, that is, training related to cliff rescue, spinal injury management, and underwater search-and-recovery using SCUBA. In addition, many lifesaving programs incorporate the use of an automated external defibrillator (AED), a device with proven effectiveness in saving drowning victims' lives before reaching the stage of biological death. In some cases, lifeguards are required to have law enforcement training and Coast Guard training, allowing the operation of motorized rescue vessels.

Introduction xv

Lawsuits alleging negligence sometimes focus on whether prompt and effective CPR was administered to a victim suffering a drowning event. In addition, administering proper spinal injury management to a victim with a suspected head or neck injury may be involved. While the Good Samaritan law protects members of the general public, it does not protect lifeguards or other aquatic professionals, including coaches, if CPR and other emergency first aid measures are not properly administered and performed in a timely manner.

CPR and the delivery of emergency first aid is often critical to the successful outcome of a patient involved in a water accident. All too often when a victim suffers a serious spinal injury from diving into shallow water or from a diving board or diving block, an improperly executed extrication and failure to properly stabilize the head and neck leads to further injury and consequently, a greater opportunity for attorneys to claim a higher level of negligence. In addition, anytime a diver shows signs of a head injury, it is the duty of the coach to perform a concussion protocol.

While rip currents have represented a hazard since humans first set foot into the ocean, new hazards have appeared on the aquatic scene, including, and to only name a few, low-head dams, shallow water blackout, medical waste, the unintentional introduction and proliferation of venomous lion fish, and the deadly brain-eating amebae, *Nigeria fowleri*. Regarding this deadly organism, evidence suggests that the microscopic organism is appearing more and more in fresh water lakes and rivers throughout America. Attorney Roger Strassburg devotes Chapter 11 to this hazard.

What makes aquatic law challenging relates to the fact that most hazards are dynamic and often hidden from the public eye. Life-threatening hazards often materialize quickly, sometimes at a moment's notice. For the purpose of an example, we will return to rip currents. Some rip currents are called "flash" rips. These rip currents are ephemeral and may sometimes form in seconds and then suddenly vanish. A school of sharks may unexpectedly appear just offshore during Labor Day weekend, making it necessary for lifeguards to close the beach for bathing; or an electrical storm may quickly materialize, leaving bathers to scurry for cover—hopefully not under a metal lifeguard tower. Sometimes river water levels rise quickly, transforming the water flow from a meandering stream to a deadly Class IV rapid. Water pollution is often responsible for sudden beach closures both in fresh water lakes and on ocean beaches. Swimming pools may quickly become contaminated from fecal matter and from the lack of proper water treatment and monitoring.

Understanding these hazards and the implicit duty about how the public must be warned is often problematic but nevertheless critically important to understand. In addition, it is important to know how people respond to warnings, and, consequently, this book devotes information relating to the effectiveness of warnings. In the case of the colored flag warnings that are common along Florida's coast and many beaches throughout the world, research discovered that flags are often ignored for several reasons and consequently not nearly as effective as believed by many aquatic experts. Note that a significant number of the American population is colorblind, so colored flags that warn about beach hazards are useless. In recent years there has been a proliferation of warning signs and because of a phenomena called "sign pollution," the efficacy of aquatic warnings is suspect. The effectiveness of hazard warnings is a subject that is often addressed as a contributing factor in complaints from attorneys claiming negligence.

Another issue making this book relevant involves the realization that rules and regulations promoting aquatic safety are inconsistent and vary from state to state. Among swimming pools, there are generally three categories of pools, including private, semi-public, and public. The highly informative book *The Encyclopedia of Swimming Pool Standards* (1999) identifies the existing and often confusing variability of swimming pool codes and regulations in every state.

The only notable exception to this ambiguity is the Virginia Grahame Baker Act, legislating anti-entrapment drains in all 50 states. Attorneys, whether defense or plaintiff, must have a thorough understanding of codes in every state because at the time of publishing this book, there is a serious lack of uniformity in regard to how each state regulates and legislates swimming pool safety.

In a recent study conducted by Fletemeyer (2015), it was discovered that swimming pools in other countries are not subject to the same regulations as pools located in the United States. Consequently, pools in foreign countries often represent accidents waiting to happen. Of the 102 foreign pools evaluated across seven different countries, all had serious problems that would have been cited for code violations in the United States. A surprising finding was that large hotel corporations based in the United States often maintained different and usually more relaxed standards in their pools located abroad than in their American counterparts.

THE ECONOMIC COST OF DROWNING

The economic cost of drowning is another important subject addressed in this book. According to the WHO, in the United States, 45 percent of drowning deaths involve individuals that are among the most economically active segment of the population. There are a number of methods of calculating the economic value of a life. The National Swimming Pool Foundation (NSPF) estimates a total cost of $5.3 billion a year in direct and indirect costs from fatal and non-fatal drownings.

Drowning causes serious economic losses in two ways—either by dying, or by being seriously injured and hospitalized. The initial costs are straightforward, including medical expenses and in the case of a fatality, the cost of funeral services. Additional costs include the loss of income and emotional loss. In a lawsuit alleging negligence, determining these additional costs becomes problematic and requires the input from an economics expert.

An emergency room visit often costs several thousand dollars, not to mention the cost of emergency transport and paramedic/EMT interventions. Sometimes emergency transport involves the costly use of a helicopter, which may exceed $20,000. In nearly every fatal drowning case, estimates are made about the economic loss to society and about the wages lost during the expected lifetime of the victim. Factoring in social, educational, demographic, and cultural variables are crucial to an accurate cost estimate.

Determining economic loss is often not only problematic but is arguably part art and part science. It is especially problematic among accidents resulting in a long-term, and in extreme cases, lifelong disabilities. In a case of a spinal injury resulting in quadriplegia, life care may cost millions, despite the likelihood of a shortened life expectancy.

A fitting conclusion to this introduction is discussion about the psychological impact of drowning, which unfortunately has never been objectively researched. However, in a paper delivered at the 2014 NSAPI Symposium in Portland by Fletemeyer, PTSD was discussed. In the past, PTSD was a condition reserved only for members of the military with experiences related to combat and death. No longer is this the case. Following interviews with teenaged lifeguards involved in fatal drowning events and family members who had lost a loved one to a fatal drowning, it was discovered that these individuals may experience a host of psychological disorders, including PTSD. Seldom are teens provided psychological counseling.

Expressions of psychological impairment are especially strong and lasting among parents whose toddler drowned due to the knowledge that drownings among toddlers are usually preventable. In these cases, as well as in other aquatic accidents, manifestations of grief and bereavement may be lasting and often require long-term psychological help.

Throughout the book, certain aquatic paradigm shifts have been recognized. For example, one involves lifeguarding practices shifting from rescue intervention to rescue prevention. In Chapter 12, Diana Sullivan identifies possibly another paradigm shift that involves an emphasis about the psychological impact of drowning. Thanks to new research focusing on this subject, although it is still in the beginning stages, not only is the economic impact of drowning considered important but so is the psychological impact.

It is important to note that there is considerable inconsistency about how states permit the opportunity for financial recovery. For example, some courts do not permit psychological costs to be included in negligence complaints. Regardless of the differences, as the principal author of this book and in regard to the legal profession, I believe that this may be a game-changer because it adds yet another factor that must be considered when developing a theory of negligence and represents another challenge that a defense attorney must be prepared to respond to. Consequently, I predict that the psychological impact on drowning will appear in increased frequency in negligence complaints and loom more significantly during mediations and in front of juries during trial.

Dr. John R. Fletemeyer

REFERENCES

Avramidis, S. (2007). The 4W Model of Drowning. *International Journal of Aquatic Research and Education*, vol. 3, no. 3, pp. 3–5.

Bittenbring, C., Bruya, L., Richiwine, M., and Youngblood, S. (1999). *The Encyclopedia of Aquatic Codes and Standards*, ed. K. Johnson. National Recreation and Park Association, vol. 1, 1–85.

Fletemeyer, J. (1999). *Introduction in Drowning, New Perspectives on Intervention and Prevention*, New York, CRC Press, pp. 1–314.

Fletemeyer, J. (2010). All American Canal Drowning Study Report. Report Submitted to San Diego Water Authority, pp. 1–44.

Fletemeyer, J. (2014). Post Traumatic Shock (PTS): American Teens Experiencing a Fatal Drowning Event. World Drowning Conference, Portland, Oregon, unpublished.

Fletemeyer, J. (2015). Overseas Swimmers Beware. *Aquatics International*, vol. 27, no. 3, pp. 27–29.

Freas, S. (1999). The History of Drowning and Resuscitation. In *Drowning, New Perspectives on Intervention and Prevention* (pp. 1–21), New York, CRC Press.

Lushine, J. B., Fletemeyer, J., and R. G. Dean. (1999). *Toward a Predictive Model for Rip Currents and Their Impact on Public Safety with Emphasis on Physical, Demographic, and Cultural Considerations in Drowning: New Perspectives in Intervention and Prevention*, ed. J. Fletemeyer, pp. 281–305. New York: CRC Press.

Principal Author

Dr. John R. Fletemeyer has a background in professional aquatics spanning thirty years. He devotes most of his academic time to drowning research and to the development of drowning prevention education programs. Dr. Fletemeyer has served as a consultant for the Centers of Disease Control and Prevention, U.S. Border Patrol, YMCA of the USA, Boy Scouts of America, and the San Diego Water Authority. He is the recipient of the Paragon Award, Judge Martin Award, the 25 Year NAUI Service Award, and the Top 100 Aquatic Professionals in America Award. He has served in several elected leadership positions, including chairman and vice-chairman of the International Swimming Hall of Fame, chairman of the National Aquatics Coalition, and president of the SE Region of the USLA. He is a currently an executive board member for the National Drowning Prevention Alliance (NDPA) and national education chairman of the United States Lifesaving Association. Dr. Fletemeyer is the author of numerous peer-reviewed articles and two books, including *Drowning: New Perspectives in Intervention and Prevention*. He has investigated over 1,000 drowning deaths and has appeared on *60 Minutes*. In addition, Dr. Fletemeyer is the founder of the Aquatic Law and Safety Institute.

Contributors

Michael D. Eriksen
Eriksen Law Firm
West Palm Beach, Florida

and

Maritime Law Association of the United States

John R. Fletemeyer
Aquatic Law and Safety Institute
Fort Lauderdale, Florida

Michael Flynn
College of Law
Nova Southeastern University
Davie, Florida

Bryan Haus
Rosenstiel School of Marine and Atmospheric Science
Miami, Florida

Julie Munger
Sierra Rescue, Inc.
and
Rescue 3 International
Truckee, California

Tim O'Brien
Former US Olympic Diving Coach

Abigail Polsby
Sierra Rescue, Inc.
Truckee, California

Lorie St John
Private Practice
Fort Lauderdale, Florida
Denver, Colorado

Roger W. Strassburg, Jr.
Private Practice
Resnick & Louis, PC
Scottsdale, Arizona
Las Vegas, Nevada
Albuquerque, New Mexico
Salt Lake City, Utah

Diana P. Sullivan
Aquatic Law and Safety Institute
Fort Lauderdale, Florida

and

Private Practice
Mental Health Services
Plantation, Florida

1 An Overview of Aquatic Law

Michael Flynn

CONTENTS

Overview and Application of Law 2
    Intentional Torts Theory 2
    Strict Liability Theory 3
    Negligence Theory 4
Other Liability Considerations in Drowning Cases 14
    Attractive Nuisance Doctrine 14
    Insurance Contract Provisions 15
    Secondary or Dry Drowning 15
Conclusion 16
Endnotes 16

Hollywood put out a comedic movie titled *A Million Ways to Die in the West*.[1] This title aptly describes the myriad ways that a person may be injured or die in an aquatic accident. The difference is that aquatic injuries are not funny! Through 2009 in the United States, there were approximately 301 million visits every year by persons over the age of six to enjoy the water at the ocean beaches, regional lakes, and swimming pools.[2] From swimming to boating to jet skiing to snorkeling to surfing, oceans, lakes, and pools provide both therapeutic and recreational activities.[3] The sobering fact is that many people sustain injuries or die from being in the water. Between 2005 and 2009 in the United States, there were an average of 3,533 fatal drownings each year.[4] Drowning is one of the leading causes of death throughout the world.[5] In the United States, children between the ages of one and four have the highest drowning rates, and most of these drownings occur in swimming pools.[6] Sadly, approximately one in five people who suffer a drowning death are children age fourteen or younger.[7] What makes these tragic statistics even more distressing is that many of these drownings were quite easily preventable.

    The purpose of this chapter is to look at one aspect of aquatic injuries, namely, drownings. In particular this chapter will examine the civil legal liability of a public or private person or entity for a drowning. This chapter will present selected court opinions that overview how the courts look at the liability for a drowning. The idea is to provide the reader with a reference and explanation, bolstered by citations, of the significant legal questions concerning liability for a drowning.

OVERVIEW AND APPLICATION OF LAW

When you review the court decisions involving drownings, the legal theories used to assign liability rest in tort law. Tort law is the subject area of law that covers private lawsuits in which the plaintiff seeks money damages from a defendant based on the defendant's fault.[8] The three key theories used to assign liability to a defendant based on the defendant's fault are *intentional torts*, *strict liability*, and *negligence*.

INTENTIONAL TORTS THEORY

The *prima facie* case for any intentional tort lawsuit requires proof of a voluntary intentional act that causes harm to a person or the person's property.[9] The key is obviously proof of intent! *Intent* is commonly defined as either a deliberate and purposeful act directed at a person that results in harm to the person or that person's property, or an act done with knowledge to a substantial certainty that the result from the act will cause harm to a person or that person's property.[10] Proof of intent makes this kind of tort very personal in kind, but proof of the motive or reason a person acted is not required or relevant to making out most intentional tort claims.[11]

In applying this simplified definition of an intentional tort to drownings, this would be the kind of circumstance where one person holds another person underwater and kills the person by drowning. This is not only a horrific and potentially criminal act, but it is also an intentional tort.

In *People v. Denis*,[12] the defendant was convicted of murder in the second degree for intentionally drowning his wife. The appellate division of the New York Supreme Court affirmed the conviction and reasoned that the evidence was sufficient to sustain the defendant's murder conviction.[13] After the defendant's wife fell from the top of a waterfall, two eyewitnesses clearly observed the defendant holding his wife's head under the water before the defendant noticed eyewitnesses coming to provide help.[14] Unfortunately it was too late; the defendant had intentionally drowned his wife.[15]

At trial, the forensic pathologist who performed the autopsy on the defendant's wife testified that the cause of death was drowning.[16] He further explained that she suffered a laryngeal spasm that was caused by water in her airway, which closed off the flow of oxygen.[17] This kind of drowning incident is referred to as a *dry drowning*.[18] Had the defendant attempted to rescue his wife immediately after running down from the cliff, she may not have suffered a drowning death. The forensic pathologist also testified that it takes about four minutes to suffer a drowning death once water obstructs a person's airway.[19] The court quickly affirmed the defendant's conviction based on the sufficient evidence available.

Thankfully, there is not a large number of these kinds of cases, but intentional tort theory remains a viable option for assigning liability in a drowning lawsuit.[20]

The primary defense to liability for an intentional tort is consent. Consent is voluntarily granting permission for the tort act to occur.[21] It seems improbable that a person would consent to being drowned. It is no surprise that there are no cases in which a court has found that consent was a valid defense in a drowning lawsuit.

STRICT LIABILITY THEORY

Strict liability is commonly referred to as "liability without fault."[22] This is perhaps a misnomer when looked at in context. The idea behind strict liability is that liability will attach if the person is engaged in an activity that presents an abnormal risk of harm.[23] Many times strict liability is applied to what courts call an abnormally dangerous or ultra-hazardous activity.[24] The courts figure out if an activity is abnormally dangerous by applying a risk–utility analysis to that particular activity.[25] The court will consider several factors in weighing the risk versus the utility of a particular activity, including likelihood and severity of the harm from the activity, the appropriateness of the location of the activity, whether the risk can be eliminated by the use of reasonable care, the value of the activity to the community, and whether the activity is a matter of common usage.[26] Any one of these factors may be enough for the court to determine that an activity is abnormally dangerous.[27] In simple terms, the court looks at the activity engaged in and determines through the risk–utility analysis if the activity itself warrants the imposition of liability for the resulting injury just because of nature of the activity. The fault, then, does not concern so much how the person acts but rather whether the activity is so dangerous that it cannot be made safe.

The earliest example of the application of strict liability in this context was to the operator of a reservoir that overflowed and damaged neighboring property.[28] The court ruled that part of the price for operating a reservoir is liability for injury and damage that results from the reservoir. The key inquiry is to analyze the activity and not the conduct involved. Perhaps the best example of a strict liability case is dynamite blasting, while the best example of an activity to which strict liability will not apply is automobile driving.[29]

The courts have been reluctant to apply strict liability to drownings. In *Wiley v. Sanders*,[30] the representative of the deceased, who drowned in a pond on the defendant's property, alleged causes of actions for negligence and strict liability against the defendant. Sanders owns a nine-acre tract of land in Louisiana, which contains a pond that he had constructed for fishing and swimming.[31] Sanders's son, Samuel, was home alone for the weekend and invited some friends over, which ended up amounting to twenty or thirty people.[32] Robert Wiley, the guest who drowned, was seen by other guests in the pond, apparently displaying no signs of difficulties.[33] It wasn't until the morning that Wiley's body was discovered in the pond.[34]

The Court of Appeal of Louisiana held that the defendant's "pond was not unreasonably dangerous and the landowner did not breach his duty to act as a reasonable person and to guard against unreasonable risk of injury or harm to guests."[35] Here, the defendant was not held liable for the drowning death that occurred in the pond on his property.[36] In reaching its conclusion, the court disagreed with the plaintiff's argument that the "totality of the circumstances of the pond" presented an unreasonably dangerous condition.[37] The pond that Sanders constructed on his property did not present any hazards or dangers to the deceased, except for the fact that the pond contained water.[38] Therefore, the defendant was not held strictly liable or negligent in causing the drowning death of the plaintiff in this case.

In *Loveland v. Orem City Corp.*,[39] the plaintiff brought causes of action against multiple defendants, including the land developer, canal operator, and the city, based on strict liability. The city escaped liability based on sovereign immunity.[40] The court held that the Loveland's alleged claim of strict liability against the defendant was inadequate, along with the other causes of actions against the defendant.[41] At the time, the Supreme Court of Utah had not applied the doctrine of strict liability to residential subdivisions.[42]

In a more recent holding out of the Fifth Circuit Court of Appeal of Louisiana, the defendant was held liable under the doctrine of strict liability, resulting from the drowning death of a fourteen-year-old boy.[43] In *Jones v. Gillen*,[44] the boy drowned in the Simalusa Creek, which was located behind the defendant campground property.[45] The creek's depth ranged from extremely shallow areas to a couple of feet, with one exception: "One area where the creek had created a hole graduating to a depth of 20 to 30 feet."[46] The defendant's campers and residents of the surrounding area used the creek and the deep-water area as a regular swimming hole.[47] The court found evidence of a cable and pulley system that was constructed across the creek at the spot where the deep hole is located.[48] Campers and residents would use the apparatus to swing across the creek and into the deep water.[49]

The court found the defendant liable under the theory of strict liability and reasoned that the swimming hole was unreasonably dangerous.[50] The swimming hole presented a danger to anyone who swam in the creek due to the lack of visibility in the water and the lack of supervision or lifeguard.[51] Further, there was evidence in the record that the defendant campground knew that someone had drowned in the swimming hole previously.[52]

Strict liability remains a viable theory of liability in drowning cases but the courts seem cautious because swimming and other water activities are such a matter of common usage. However, in situations like the *Jones* case, where there was proof that the campground operator knew that someone else had drowned in the area and did not take any steps to correct the danger, strict liability works well to hold the defendant liable.

The defenses to strict liability are, for the most part, the same as the defenses to negligence, which will be covered later in this chapter.

NEGLIGENCE THEORY

By far the most widely used theory of liability in drowning cases is negligence. A *prima facie* case of negligence requires proof of four elements: (1) the existence of a duty to exercise reasonable care; (2) breach of that duty to exercise reasonable care; (3) a causal connection, namely that the breach of the duty to exercise reasonable care is the cause in fact and proximate or legal cause of injury; and (4) injury in fact.[53]

Drowning is proof of injury in fact. But the fact of injury does not prove liability for negligence. Rather, all four *prima facie* elements must be proven.

The existence of duty to exercise reasonable care is a question of law and arises from an analysis of the risk of harm of engaging in the act.[54] The courts have ruled that the existence of a professional relationship is enough to impose by law a duty to exercise reasonable care.[55] For example, the doctor–patient or lawyer–client relationship.

An Overview of Aquatic Law 5

In other instances, the imposition of a duty to exercise reasonable care may be set by statute or ordinance,[56] for example, traffic laws. Still, in other instances, the existence of a duty to exercise reasonable care depends on the status of the person.[57] For example, an undiscovered trespasser does not require a duty to exercise reasonable care, but a store customer or other invited person is entitled by virtue of that status to be protected by a duty to exercise reasonable care.

In most other cases, the courts have used a formulaic approach in determining the existence of a duty to exercise reasonable care by evaluating the likelihood and severity of the harm when engaging in the act versus the cost or burden to prevent the harm.[58] If the risk and severity of the harm is greater than the cost to prevent the harm, the activity is viewed as creating an unreasonable risk for which the courts will impose a duty to exercise reasonable care. Conversely, if the risk and severity of the harm is less than the cost to prevent the harm, the activity is viewed as creating a reasonable risk of harm for which the court will not impose a duty to exercise reasonable care. In the context of drowning cases, the risk and the severity of harm of drowning when engaged in water activities is a constant, while the cost of prevention varies with the circumstances.

Breach of the duty to exercise reasonable care is more of a question of fact. In this instance the courts will look objectively at the acts to determine if the conduct of the person conforms to what a reasonably prudent person would do or not do under the same or similar circumstances.[59] The key is that this determination is not subjective but is informed by the facts surrounding the activity.[60] In the context of drowning cases, this is where the failure to post properly trained lifeguards or to adequately warn of dangerous water hazards may signal a breach of the duty to exercise reasonable care.

The third element of the *prima facie* case of negligence is causation. Causation requires proof of both cause in fact and proximate or legal cause. Cause in fact is exactly how it sounds—a question of fact that requires proof of cause and effect.[61] The courts ask: If but for the breach of the duty to exercise reasonable care, would the person have been injured?[62] This construct can be applied to single or multiple persons acting alone or in concert.[63] The courts have even gone so far as to broaden this construct to conclude that if the breach of duty multiplied or contributed to the person's injury, then the actor is a cause in fact of the injury.[64] The key for the trier of fact, the court or a jury, is to determine that the breach of duty is a substantial factor in the person being injured. As to drowning cases, if the activity engaged in contributed or solely resulted in a drowning, cause in fact would be proven.

As to the requirement of proximate or legal cause, the inquiry does not really deal with causation at all. Rather, proximate cause is the construct that limits the reach of liability for negligence and is most often also a question of fact for the trier of fact.[65] The guideline for determining proximate cause is fairness. With this in mind, the courts ask the question if the injury sustained by this person is reasonably foreseeable from the breach of duty.[66] In making this determination the courts can look at several factors, including remoteness in time, space, and difference; the presence or absence of intervening acts; the zone of the risk created by the breach of duty; and the practical social, economic, and political public policies implicated by any decision.[67] As to drowning cases, all of these factors can weigh in favor of or against finding proximate cause. For example, a person who drowns at night when a beach or lake or pool is not open may weigh in favor of not finding proximate cause as to the act. Meanwhile, a person who drowns during the day when a beach or lake or pool is open may weigh in favor of finding proximate cause.

In both negligence-based and strict liability–based lawsuits, the defenses of contributory negligence and assumption of the risk have been successful in limiting or denying liability for drownings. Contributory negligence means negligence on the part of the plaintiff, the injured person.[68] The construct and requirements of an ordinary negligence claim apply equally to prove contributory negligence. In the context of a drowning case, a person who is not an expert swimmer but who attempts to swim too far from shore or in too-deep water may be found to be contributorily negligent. The key to contributory negligence is that the court looks at the conduct of the injured person to see if that person possessed a duty to exercise reasonable care, breached said duty to exercise reasonable care, that the breach of duty at least contributed to the injury suffered by the person, that the injury was foreseeable, and finally, that that the person suffered injury—in this case, drowning.[69]

How the courts handle contributory negligence depends on where you live. There are a small number of states that have ruled that contributory negligence bars any recovery for the injured person.[70] This rule absolves the negligent actor of liability. The majority of states view this rule as too harsh. Instead, the majority of states have ruled that contributory negligence reduces the injured person's recovery by the percentage of the injured party's fault.[71] Some other states only permit contributory negligence to bar an injured person's recovery if the injured person's percentage of fault is above 50 percent or is greater than the percentage of fault assigned to the negligent actor.[72] Although not as harsh a rule, this construct, referred to as *comparative negligence*, requires the trier of fact to determine percentages of fault, which can be a difficult decision.

The other popular defense is assumption of the risk. Assumption of the risk is the counterpart to the defense of consent in an intentional tort case. Assumption of the risk means that the injured person subjectively knew of the risk and the magnitude of the risk before engaging in the activity and then voluntarily chose to engage in the activity anyway.[73] In simple terms, they consented to the activity. In the context of a drowning case, proof that the injured person subjectively knew that rip currents were present in the ocean and that rip currents can lead to drowning but then voluntarily chose to swim in that area of water can be seen as assumption of the risk. Assumption of the risk, just like contributory negligence, can be treated as a complete bar to recovery or a percentage reduction of the amount of any recovery by an injured person, depending on the state you live in.

As incongruous as it may seem, the defense of contributory negligence does apply in strict liability–based claims as well.[74] It follows that assumption of the risk would apply to a strict liability–based case, but not a negligence-based defense such as contributory negligence. The reason for the extension of the application of contributory negligence to strict liability is the notion that it is fairer to have all parties be liable based on the percentage of fault. In the context of a drowning case, proof that the injured person knew that the pool was not suitable for diving and dove anyway may

An Overview of Aquatic Law 7

not bar the injured party from recovery but may reduce any recovery by the percentage of the injured person's fault.

The courts have applied negligence theory to a myriad of drowning cases. In *Rickwalt v. Richfield Lakes Corp.*,[75] the plaintiff drowned while vacationing at a private beach resort in Michigan.[76] The defendant staffed the beach with lifeguards, but the sixty-six-year-old plaintiff, who went swimming with two of his grandchildren, drowned despite the presence of lifeguards and ten to fifteen other resort guests.[77] The plaintiff's body was found floating less than fifteen feet from shore directly in front of a lifeguard station.[78] The court had little trouble in finding the defendant beach resort liable for the plaintiff's death because of the failure of the lifeguards employed by the defendant to adequately monitor the beach area and respond to the plaintiff while in distress.[79] Therefore, the negligence on the part of the lifeguards employed or otherwise retained by a private beach owner will result in liability being attached to the defendant owner of a private beach resort.[80]

The following case involves the liability of a public entity, Metropolitan Dade County, for a drowning.[81] In *Metropolitan Dade County v. Zapata*,[82] the decedent and some friends went to the defendant's Crandon Beach Park, which was supervised by four lifeguards at the time. The decedent and two others walked in shallow water out to a sandbar that was located about seventy-five yards off shore.[83] While returning to shore, the boys accidently stepped into a six-to-eight-foot drop-off and one of the boys unfortunately drowned within minutes.[84] The defendant's beach was staffed with lifeguards, but they did not notice or attempt to rescue the plaintiff.[85] The plaintiff prevailed in the Circuit Court, with a jury verdict finding that defendant, Metropolitan Dade County, was 90 percent liable to the plaintiff regarding the wrongful death of the deceased.[86] Therefore, this case demonstrates that a public beach can be held liable for drowning deaths at beaches staffed with lifeguards.

On the other hand, there are cases in which public entities or beaches are not held liable for alleged negligent acts that resulted in drowning deaths at such locations.[87] In *Suarez v. City of Texas*,[88] two nine-year-old twins went into the ocean upon arriving to the beach with their parents.[89] Shortly thereafter, the twin girls were struggling to remain afloat as a strong current swept them out.[90] The girls' father attempted to rescue his daughters, but the current was too strong and all three guests drowned.[91] Although the beach was not equipped with signs prohibiting swimming or warning of the swimming dangers at the public beach, the court did not find the City of Texas liable for the three plaintiffs' drowning deaths.[92] The court reasoned that the city was immune from liability based on the protections afforded public entities by the applicable recreational use statute.[93]

Similarly, the state of New Hampshire enacted a recreational use statute that limits landowner liability by absolving liability for personal injury or property damage that is not caused intentionally by the landowner.[94] In *Coan v. New Hampshire Dept. of Environmental Services*,[95] the defendant was not held liable for the plaintiffs' drowning deaths because the Department of Environmental Services was immune from liability under the recreational use statute. The defendant operated an electricity generating plant, which was connected to the dam on the lake.[96] The dam created a dangerous flow coming out of the dam into the lake, which made the currents deadly in the area where the plaintiffs were swimming.[97] However, the aforementioned recreational use statute provided full immunity to the defendant for the liability of the plaintiffs' drowning deaths.

A brief explanation of the purpose and impact of recreational use statutes is appropriate at this point in the chapter. Recreational use statutes provide public and private beach owners with immunity from liability for drowning in certain circumstances. These recreational use statutes, enacted by state legislatures, encourage public and private landowners to permit people or guests to enter their land for the purpose of engaging in recreational activities.[98] For example, like the previously mentioned New Hampshire recreational use statute,[99] the California recreational use statute establishes the limited liability of a private landowner who provides people with the ability to use the land for recreational purposes.[100] This statute provides an exception from the general rule that a landowner owes a duty of reasonable care to anyone entering the land.[101] Thus, recreational use statutes will be upheld when the statute is intended to protect the defendant landowner from liability for injury resulting from recreational activities on their land. If the statute is not intended to protect the liability in dispute, the recreational use statute may be found inapplicable by the courts, thereby imposing liability on the defendant.

Recreational use statutes do not provide a general immunization for public and private beach, lake, and pool owners. It is a common element throughout the various recreational use statutes that if the landowner grants the public access to his or her land without a fee, the landowner will not be held liable for deaths that occur on his or her land, including drownings.

Courts have even absolved landowners of liability resulting from drowning deaths where the landowner charged a fee based on a per-vehicle parking.[102] In *Cole v. South Carolina Electric and Gas, Inc.*,[103] a patron drowned while visiting the defendant's recreational beach area. The court held that the applicable recreational use statute operated to bar the plaintiff's family from recovering based on allegations of negligence.[104] The court reasoned that the parking fee was not considered to be a "charge" as defined by the state's recreational use statute.[105] Only the driver of the plaintiff's party paid the $3 parking fee.[106] The fee, as defined by statute, refers to an individual, per-person fee that is charged upon entering the landowner's property. The parking fee in this case is clearly distinguishable from a per-person entrance fee. Here, access to the defendant's recreational beach area was free to patrons arriving by foot, by bike, or by swimming to the site.[107]

In the alternative, liability would attach to a landowner who does not fall within the statutory meaning of the states' recreational use statute. In *Rivera v. Philadelphia Theological Seminary of St. Charles Borromeo, Inc.*,[108] the plaintiff drowned in the defendant's indoor pool. The court held that Pennsylvania's recreational use statute did not extend to provide immunity to the defendant's indoor pool, therefore finding the defendant liable for the drowning death.[109]

Even with the seemingly broad grant of immunity given to some private and public landowners through the recreational use statutes, many operators of water areas use lifeguard and warning signs to try to protect people who frequent water areas.[110] In *Dontas v. City of New York*,[111] the plaintiff was sixteen years old when he drowned during swimming class at the high school he attended in New York.[112] The boy was discovered floating in the pool shortly after the swimming class had ended.[113] The defendant's swimming instructor attempted to apply artificial respiration to the plaintiff immediately after removing the plaintiff from the pool; however, it was to no avail.[114] The court ruled in favor of the plaintiff and awarded millions of dollars in damages for the negligence of the public entity defendant.[115]

In *Ward v. City of Millen*,[116] the court held a public defendant liable for a drowning death that occurred at a pool leased by the city to an association who provided lifeguards. The defendant was alleged to have been operating a pool with the water in a murky condition, thereby preventing the lifeguards or pool guests from clearly seeing the pool's depth.[117] The deceased was standing in about four feet of water when he was first observed in the pool.[118] However, the single lifeguard on duty did not observe when the decedent went below the surface of the murky water.[119] In fact, the decedent was not discovered until the lifeguard sent someone into the pool to search the bottom.[120] Here, the defendant lifeguard and lessee of the city-owned pool clearly failed to maintain the pool in a condition satisfactory under the law by allowing the guests to remain in the pool with murky water.[121]

Liability may also be imposed for the negligence of defendants who are private entity pool owners when a drowning death occurs at such pool.[122] In *Coleman v. Shaw*,[123] the defendant motel owner was held liable for the death of a patron who drowned in the motel's unsupervised pool. In this scenario, South Carolina had a statute in place that regulated the requirements related to swimming pool safety precautions.[124] The plaintiff's niece observed the plaintiff experiencing distress in the pool, whereupon the plaintiff's body sunk and then was floating in the deep end of the pool.[125] The niece alerted the plaintiff's wife, who was sleeping in the room at the time of the incident.[126] After many requests for assistance, the motel front desk attendant would not help the plaintiff's wife.[127] The plaintiff's body remained at the bottom of the pool for a prolonged period of time until rescue workers arrived on scene.[128] The court held that the defendant's violation of the statute was negligence per se; thus, the defendant motel was negligent in failing to abide by the applicable statutory law.[129] The motel pool was not staffed with an employee trained in first aid care and it did not have any emergency safety equipment readily accessible at the pool.[130] Therefore, the defendant's violation of the statute constituted the defendant's breach of duty of care owed to its patrons, including the plaintiff in this case.

However, in *Odom v. Lee*,[131] the court did not find the defendant pool owner liable for the drowning death that occurred on its premises. A six-year-old boy drowned in the defendant's swimming pool when the defendant's family was not home.[132] The court held that the defendant was not liable for the drowning death of the plaintiff because the pool was surrounded by a padlocked chain fence and the defendant's family was not at the house when the incident occurred.[133] Although the plaintiff had frequently played with the defendant's child, in this instance, the defendant could not be held liable because the plaintiff was not at the defendant's house when the defendant's family left or earlier that day before the family left.[134] Therefore, without a causal connection linking the defendant landowner to the drowning death of the plaintiff, the court could not impose liability on the landowner.[135]

In the following example, liability is still imposed on the defendant private pool owner although the victim suffered what is called a near-drowning incident,[136] eventually leading to the victim's death. An eleven-year-old girl was on a fifth-grade class trip to the YMCA aquatic facility in Brandywine.[137] The plaintiff was discovered floating face down in the pool by the defendant's lifeguards.[138] When the lifeguards became aware of the situation, the plaintiff was removed from the pool, where she was responsive after vomiting water.[139] The plaintiff was rushed to the hospital for treatment and eventually died due to complications of the near-drowning incident.[140] The defendant, the YMCA, was held liable for the death of the eleven-year-old decedent at the conclusion of the jury trial.[141] In this scenario, the defendant lifeguards had not been properly monitoring or guarding the subject pool, otherwise the lifeguards would have noticed the plaintiff in distress.[142]

The property owner of land containing a lake can be either a public or a private owner. The numerous lakes across the country are either natural or manmade, ranging in sizes and depths. Lakes provide a location for aquatic-related activities, creating many issues of liability for potential defendants.[143] The following cases will examine the potential liability for drowning deaths that occur in lakes.

In *Simons v. Hedges*,[144] an employee at a resort hotel, drowned while using the employer's canoe for recreational purposes while off-duty. The resort hotel was located on Blue Mountain Lake in the Adirondack Mountains.[145] During an afternoon off-period, the employee used one of the employer's canoes for recreational purposes on the lake.[146] Unfortunately, the canoe capsized and the employee drowned in a part of the lake owned by the defendant employer.[147] The court held that the defendant was liable for the drowning death of its employee because the defendant permitted the employees to use the resort facilities for recreational purposes.[148] Further, the court reasoned that the defendant induced the employees to use the resort's facilities as part of the contract of employment.[149] Therefore, a defendant can be held liable in an employment relationship where an employee suffers death by drowning from engaging in aquatic activities on the defendant's lake.

In a situation similar to the aforementioned case, a laborer at a defendant's camp drowned while swimming in the defendant's lake.[150] However, *Leonard v. Peoples Camp Corp.*[151] can be distinguished from *Simons v. Hedges* because the lake in this case was completely surrounded by the defendant's land, which was approximately 285 acres.[152] The employees had never been prohibited from using the lake and customarily used the lake for fishing on their days off.[153] The court held the defendant camp liable for the drowning death of its employee because the death occurred in the course of the decedent's employment.[154] Swimming was found to be a normal and reasonable activity of the defendant's employees at the camp. Thus, the defendant could not avoid liability by arguing that the decedent went swimming in the lake on the decedent's day off from work.[155]

On the other hand, there are situations where the defendant may not be found liable for a drowning death that occurred in the defendant's lake.[156] In *Blacka v. James*,[157] the defendant owned and operated a three-acre lake and charged $5 for admission by the general public.[158] The plaintiff was observed by his friend playing on the recreational platforms in the center of the lake, but could not be found when it was time to leave.[159] The defendant's lifeguards began searching the premises and eventually discovered the decedent's body in the roped-off section of the lake, about ten feet away from the platform that the decedent was playing on.[160] The court did not hold the defendant landowner liable for the death of the young boy because there was no evidence or indication by any of the few hundred people in or around the lake that the plaintiff was in distress or possibly drowning.[161] Therefore, liability could not be imposed on the defendant who owned the lake and staffed the lake with lifeguards.

The following case involves the tragic drowning deaths of a mother and her children in a lake undergoing a government project, where the defendant was not found liable for negligence in failing to warn of the dangerous conditions present in the lake.[162] In *Hutchinson v. Township of Portage*,[163] a dredging project took place around the boat launch area of the lake, supervised by a government employee. The defendants dredged to a depth that was beyond what the respective permit called for.[164] The court did not hold the defendant government supervisor liable for failing to post warning signs of the depth as a result of the dredging work because the government employee was protected by sovereign immunity as set forth in Michigan statutory law.[165] The court reasoned that drop-offs can naturally occur in lakes and further stated that the parents should have been closely supervising the young children because the children had little or no swimming skills.[166]

The next set of cases involves drowning deaths that occurred in rivers and other bodies of water. The first case example involves a public entity being held liable for the drowning deaths of two victims. In *Trinity River Authority v. Williams*,[167] the two plaintiffs drowned while fishing in the Trinity River.[168] The plaintiffs put their boat into the river just outside the restricted area near the dam.[169] Strong backcurrents caught the plaintiffs' boat and drew it up into the restricted area located near the dam, causing the boat to capsize.[170] Both plaintiffs drowned as a result of the dangerous backcurrents.[171] Liability was alleged against the City of Houston and the Trinity River Authority (TRA) for failure to give adequate warning of the dangerous backcurrents in the river just below the restricted area.[172] Defendant TRA claimed governmental immunity based on the floodgate provision of the Texas Tort Claims Act.[173] However, the court held that the provision did not apply because the plaintiffs did not allege the negligent operation of the floodgates.[174] Further, the court held that the plaintiffs were guilty of contributory negligence because they knew or should have known about the strong backcurrents just below the restricted area.[175] The finding of contributory negligence did not completely bar the plaintiffs from recovering damages; it reduced the damages by a percentage determined by the jury.[176]

On the other hand, in a similar situation, the private-entity defendants were not held liable for the drowning deaths of two young boys.[177] In *White v. Georgia Power Co.*,[178] two boys, ages nine and twelve, drowned in the Oconee River about four miles downstream from a dam being operated by the defendant Georgia Power Company.[179] The court held that the boys assumed the known risk of drowning when they entered the river and were immediately swept away by the swift current, especially because the boys did not swim.[180] Additionally, the court held that the defendants were under no obligation to erect warning signs or place barricades in the river.[181] Therefore, public and private entity defendants can avoid liability for a drowning death that occurs in its river or body of water and do not have to warn of open and obvious dangers, such as the swift current caused by the dam turbines in this case.[182]

The following case involves the liability of an employer for the drowning death of its employee that occurred in the course of employment. In *Webb v. Hunter*,[183] the plaintiff was working on stream banks that ran under roadway bridges.[184] The stream's depth varied and the plaintiff had to get into the stream in order to complete cleaning debris from the stream slope near the blow hole.[185] In a matter of seconds, the other employees heard the plaintiff calling for help, but then he disappeared from sight and subsequently drowned.[186] The court held the defendant employer liable for the drowning death of its employee because it had occurred in the course of employment and not in the course of a personal frolic.[187] When the plaintiff drowned in the deep hole of the stream, he was undisputedly performing within the course of his employment duties.[188] Therefore, liability may be imposed on a defendant who does not own the body of water, but is merely performing work.

The courts have applied both the defense of contributory negligence and assumption of risk liberally in drowning cases. In *McFarland v. Grau*,[189] a fourteen-year-old boy and his friends paid admission at a private lake to access the lake for swimming, fishing, or recreational activities. The plaintiff and his friends knew that the plaintiff was not a strong swimmer when they decided to swim across the lake to the deeper side.[190] The plaintiff expressed that he didn't think he could swim across the width of the lake, but his friends began swimming across.[191] However, the plaintiff changed his mind and attempted to swim across the lake, but he could not make it and screamed for help.[192] Unfortunately, the plaintiff's friends did not respond in a timely manner and the plaintiff drowned.[193] The plaintiff alleged that the defendant landowner was liable for the plaintiff's drowning death due to the negligence in failing to take any precautions for the safety of the defendant's patrons.[194] The defendant virtually conceded negligence in some aspects; however, the defendant asserted that the plaintiff was negligent in contributing to his own death.[195]

The court stated that the defendant did not employ any lifeguards or post any warning signs indicating the depth of the lake.[196] However, the court did not find a causal connection between the defendant's lack of safety equipment and warnings and the plaintiff's drowning death.[197] The court held that the plaintiff was contributory negligent because he possessed the knowledge and appreciation of the danger involved in swimming across the lake.[198] The plaintiff fully understood the danger of swimming in deep water as manifested by the plaintiff when he told his friends that he didn't think he could swim across the lake.[199] Therefore, the court held that the defendant was absolved from liability in this case because the plaintiff's negligent conduct barred the plaintiff from recovering for his drowning death.[200] Further, the court reasoned that the age of the plaintiff was not a relevant issue because the evidence shows that the plaintiff was a normal, intelligent, and industrious young man.[201] Thus, negligent landowners who engage in recreational aquatic activities may escape or at least reduce any liability in the circumstances where the plaintiff was also negligent in causing injury or death to that plaintiff.

Both defenses of contributory negligence and assumption of the risk were raised in *Garber v. Prudential Ins. Co. of Am.*[202] A photographer drowned while employed to take undersea photographs for the corporation.[203] The court was reluctant to allow the jury to hear the affirmative defense of assumption of the risk because the photographer was inexperienced with diving and the heavy equipment associated with the sport.[204] The court found the defense of assumption of the risk to be without merit but allowed the contributory negligent defense.[205] The photographer knew that he had insufficient training before attempting to take photographs while diving in the ocean.[206] Therefore, it was proper for the jury to hear the question of contributory negligence in this case.

In *Metropolitan Dade County v. Zapata*,[207] the deceased drowned at a public beach while returning to shore after going out to a sandbar. The county claimed that the trial court erred in precluding any argument or comment relating to the phrase "assumption of the risk."[208] The Third District Court of Appeal of Florida agreed with the defendant, a public entity in this case, that it was an error to prohibit the defense's argument from using the phrase "assumption of risk."[209]

Courts have applied contributory negligence to minors in some situations, taking the specific facts of each case into consideration with the knowledge and experience of the child. In *Carter v. Boys' Club of Greater Kansas City*,[210] the plaintiff was a twelve-year-old child who did not know how to swim. The child attended the club's pool for his first free-swim session with about ten other young boys.[211] The boy drowned in the deep end of the pool without anyone noticing him in distress.[212] The court reasoned that the boy's contributory negligence was based upon the fact that he had the knowledge and appreciation of the danger and risk associated with his conduct.[213]

The following cases involve situations where defendants may escape liability for the death or injuries resulting from patrons diving into pools. In *Leakas v. Columbia Country Club*,[214] the plaintiff was severely injured when he dove into the defendant's pool in the shallow end and struck the bottom. The court held that the defendant country club was not liable for the severe injuries sustained by the plaintiff because the plaintiff was either contributory negligent or assumed the risk of injury by improperly diving into the defendant's pool.[215] Liability is not imposed in this case because the plaintiff observed the area from where he dove into the pool, which was clearly shallow.[216] The court reasoned that if the plaintiff had sufficiently surveyed the area and inquired as to whether it was safe to dive from the shallow end, then the plaintiff should have seen the depth marker.[217] If the plaintiff had dove from the defendant's diving board, located in the deep end of the pool, and still managed to sustain severe injuries resulting from striking the bottom of the pool, then the defendant may be held liable. However, when a defendant's pool guest takes it upon him or herself to dive into an improper area of the pool, the defendant may avoid liability entirely if the court finds the plaintiff contributory negligent or that the plaintiff assumed the risk of the injuries sustained by that plaintiff's actions.

In addition, in *Kalm, Inc. v. Hawley*[218] the defendant was not held liable for injuries sustained by a patron who dove into the pool and struck the sloped bottom. The defendant's pool was equipped with a diving board and employed a lifeguard.[219] On the day of the incident, the plaintiff watched his daughter receive some diving instructions or tips from the defendant's lifeguard at the pool.[220] Later that evening, the plaintiff's daughter asked him to demonstrate the diving technique that the lifeguard had showed his daughter earlier.[221] The plaintiff dove into the pool from one of the sides, not from the diving board, and sustained injuries to his head as it struck the sloped bottom of the pool.[222] The court did not find the defendant negligent for the plaintiff's severe injuries because the plaintiff brought about his own injuries.[223] The court reasoned that the plaintiff was negligent and contributed to causing the injuries by knowingly diving into the pool from an inappropriate area.[224]

OTHER LIABILITY CONSIDERATIONS IN DROWNING CASES

The previous discussions in this chapter has chronicled and indexed the basic theories of liability employed in drowning cases. There are a number of other legal considerations and applications that impact drowning lawsuits—some familiar and others not so familiar.

ATTRACTIVE NUISANCE DOCTRINE

A defendant may use the elements of the attractive nuisance doctrine to avoid liability. For example, in *Saga Bay Prop. Owners Ass'n v. Askew*,[225] the defendant was not found liable for the drowning death of a child that occurred in an artificial lake on the defendant's property. The court reasoned that absent some unusual danger or dangerous condition, the defendant is not liable for a drowning death, which is foreseeable in any body of water.[226]

Although one side of the lake was developed with a sandy beach for use by the general public, the side in which the child drowned contained typical plant life and construction debris.[227] The court acknowledged that the sudden change in depth on the side where the drowning occurred is characteristic of a lake and did not create an unreasonable danger to provide a basis for owner liability.[228] The defendant was able to use the attractive nuisance doctrine to avoid liability in proving that the artificial lake on the defendant's property did not create an unusual dangerous condition. The attractive nuisance doctrine holds owners and occupiers of land, including beach, lake, and pool areas, liable for the injury to trespassing children where the landowner knows or has reason to believe that the condition on the land, which presents a particular and unusual risk of injury, makes it likely that children will trespass.[229]

Landowners may also avoid liability in an event where the river, pond, reservoir, or other body of water is found not to be an attractive nuisance following a drowning death in the landowner's body of water. In *Smith v. Chicago & E.I.R. Co.*,[230] a nineteen-month-old baby drowned in an abandoned reservoir on the defendant's property.[231] The court relied on the majority rule from Illinois in holding that no actionable attractive nuisance existed on the defendant's property so as to place liability upon the defendant.[232] Therefore, the abandoned reservoir on the defendant's premises did not need to be maintained or blocked off because the owner of private land is under no obligation to protect the safety of strangers who may wander onto the defendant's property—in this case, a nineteen-month-old baby who drowned.

The attractive nuisance doctrine, as mentioned previously, applies to things or conditions artificially created and maintained by a landowner.[233] When children become attracted to the artificial condition and their recreation results in death by drowning, liability may attach to the landowner who created and maintained such artificial condition. However, a landowner may not owe a duty to keep children off of his or her property that contains an artificial aquatic condition if there is no unusual element of danger.[234]

INSURANCE CONTRACT PROVISIONS

Many insurance policies, including life insurance policies, contain exceptions or exclusions, with one in particular being the aviation exclusion clause. In theory, the aviation exclusion clause prevents the decedent's representatives from recovering what would normally be a recoverable loss under the given insurance policy.[235] In fortunate situations when survivors of an aircraft carrier that makes a crash landing into the ocean are rescued, the airplane exclusion clause will not apply. However, when the survivors of an aircraft carrier that makes a crash landing into the ocean who are not rescued and do not survive, the decedent's family members are barred from recovering losses because the cause of death was ruled to be a result of the airplane crash. The next case involves an airplane that crashes into the ocean, and the survivors of the crash escape the wreckage and subsequently drown sometime thereafter.[236]

The airplane exclusion clause was applicable in *Howard v. Equitable Life Assurance Soc. of U.S.*,[237] thus barring the plaintiff from recovering under the insurance policy.[238] The aircraft experienced mechanical difficulties over the Atlantic Ocean, causing the plane to crash into the ocean.[239] The plaintiff and others onboard the aircraft survived the crash and remained in the ocean throughout the night, floating with the assistance of life jackets.[240] Unfortunately, water seeped into the plaintiff's life jacket and he drowned before the other survivors were rescued the following morning.[241] The court applied the airplane exclusion clause in the plaintiff's insurance policy and held that the plaintiff's legal cause of death resulted from the airplane crash.[242] Therefore, the plaintiff was barred from recovering damages under the policy, although the plaintiff had survived the crash and remained floating in the ocean for a period of time.

Alternatively, the airplane exclusion clause will be held inapplicable by courts if the evidence cannot show whether the plaintiff's death resulted directly or indirectly from injuries sustained during the aircraft crash.[243]

SECONDARY OR DRY DROWNING

The issue of secondary and dry drowning also presents an interesting topic. Generally, drowning incidents involve situations in which the victim's lungs fill with water.[244] On the other hand, there are instances where the drowning victim suffered secondary drowning, or dry drowning, a rather new focus or topic under the law.

In situations where there are lifeguards or supervision present, the plaintiff will make the argument that if lifeguards had been actively monitoring and carrying out their responsibilities, they would have had time to respond to the plaintiff in distress, potentially preventing death or minimizing injury to the plaintiff.

In *Corda v. Brook Valley Enterprises, Inc.*,[245] the plaintiff drowned in the defendant country club's pool. The lifeguard left his station to put some chairs and umbrellas away while the plaintiff was in the swimming pool.[246] When the lifeguard returned, he noticed a bluish-gray object in the pool below the surface.[247] The lifeguard removed the plaintiff's body from the pool and attempted to revive the plaintiff by mouth-to-mouth resuscitation, but was unsuccessful.[248] The court found that there was some evidence that the plaintiff died from a dry drowning.[249] A dry drowning incident can be distinguished from a wet drowning incident because the voice box becomes obstructed, which prevents water and air from entering the person's lungs.[250] However, the defendants contend here that if the plaintiff suffered a dry drowning death, then the resuscitation would have been ineffective; thus, the failure to remove the plaintiff's body from the pool would not have been the legal cause of the plaintiff's death.[251]

Secondary or delayed drowning are other terms for the aforementioned dry drowning phenomenon. There is a case that presents the liability of a school district for a drowning incident that took place in their aquatic center. In *Spady v. Bethlehem Area School Dist.*,[252] the plaintiff was a fifteen-year-old student attending Liberty High School. The plaintiff complained that he was not feeling well during the PE swimming class, but he participated in the swimming lessons anyway.[253] Following the swim class, the plaintiff went to English class, where the symptoms of secondary or delayed drowning arose.[254] The teacher heard a noise and noticed the plaintiff's head was down on the desk, and the plaintiff appeared to be having a seizure.[255] Additionally, a pink, frothy substance was coming out of the plaintiff's nose and mouth.[256] The school nurses attempted cardiopulmonary resuscitation and administered two electric shocks to the plaintiff's chest with an automated external defibrillator.[257] Emergency responders rushed the plaintiff to the emergency room but he died shortly thereafter at the hospital, as a result of secondary or delayed drowning.[258]

At trial, the plaintiff's expert opined that the plaintiff died as a result of delayed drowning due to a chemical-related toxicity from the pool chlorine.[259] The public entity defendant here, the school district, did not provide adequate training to its physical education teachers because the subject swim instructor did not know anything about delayed or secondary drowning when the plaintiff died.[260] Further, the court indicated that the defendant's swim instructor did not know what symptoms to look for in scenarios with secondary or delayed drowning.[261]

Cases involving secondary, delayed, or dry drowning incidents are relatively new terms as discussed in a legal sense. Landowners or property owners who engage in aquatic-related activities may be held liable for a drowning death as described in this section. The lack of knowledge or training to understand the symptoms and effects of secondary, delayed, or dry drowning is not an excuse in the eye of the law to avoid liability for negligence.

CONCLUSION

The foregoing materials by no means cover all of the legal principles and case precedents that govern the legal liability for a drowning. However, this material should provide you with the basics concerning the legal theories used in drowning cases and some case law to help you in the search for the best argument and best precedent in the work up of a drowning case.

ENDNOTES

1. *A Million Ways to Die in the West*, released May 30, 2014.
2. U.S. Census Bureau. 2009. Statistical abstract of the United States: 2012. Arts, recreation, and travel: Participation in selected sports activities, 2009

3 See 47 ALR 4th 262 (originally published in 1986).
4 Laosee, OC, Gilchrist, J, Rudd, R. 2012. Drowning 2005–2009. *MMWR*, 61(19):344–347.
5 *Id.*
6 Centers for Disease Control and Prevention, National Center for Injury Prevention and Control. Web-based Injury Statistics Query and Reporting System (WISQARS) [online]. Cited May 3, 2012. Available from http://www.cdc.gov/injury/wisqars.
7 Laosee, OC, Gilchrist, J, Rudd, R. 2012. Drowning 2005–2009. *MMWR*, 61(19):344–347.
8 Dobbs, DB, Hayden, PT, Bublick EM. 2011. *The Law of Torts* (2nd ed.), West, St. Paul, MN, § 1.
9 *Id.* at § 29.
10 *Id.*
11 *Id.* at § 29.
12 *People v. Denis,* 276 A.D.2d 237, 716 N.Y.S.2d 718 (2000).
13 *Id.*
14 *Id.* at 241.
15 *Id.* at 239.
16 *Id.* at 240.
17 *Id.*
18 *Id.*
19 *Id.*
20 See *Elkin v. State*, 531 So. 2d 219 (Fla. Dist. Ct. App. 1988); *Com. v. Carrasquillo*, 82 Mass. App. Ct. 1107, 970 N.E.2d 814 (2012); *State v. Chappell*, 225 Ariz. 229, 236 P.3d 1176 (2010); *State v. Hayes*, 347 S.W.3d 676 (Mo. Ct. App. 2011); Vaughn v. State, 261 Ga. 686, 410 S.E.2d 108 (1991); *State v. Willoughby*, 58 N.C. App. 746, 294 S.E.2d 407 (1982); *Davidson v. State*, 558 N.E.2d 1077 (Ind. 1990); *State v. Williams*, 652 S.W.2d 102 (Mo. 1983); *Conley v. State*, 790 So. 2d 773 (Miss. 2001); *State v. Matthews*, 450 So. 2d 644 (La. 1984); *State v. Libby*, 546 A.2d 444 (Me. 1988).
21 Dobbs, DB, Hayden, PT, Bublick EM. 2011. *The Law of Torts* (2nd ed.), West, St. Paul, MN, § 34.
22 *Id.* at § 437.
23 *Id.* at § 441.
24 *Id.* at § 443.
25 Restatement (Second) of Torts § 520 (1977).
26 *Id.*
27 *Id.*
28 *Rylands v. Fletcher*, LR 3 HL 330 (1868).
29 Dobbs, DB, Hayden, PT, Bublick EM. 2011. *The Law of Torts* (2nd ed.), West, St. Paul, MN, § 443.
30 *Wiley v. Sanders*, 37,077 (La. App. 2 Cir. 6/13/03), 850 So. 2d 771, *writ denied*, 2003–1986 (La. 10/31/03), 857 So. 2d 487.
31 *Id.* at 773.
32 *Id.*
33 *Id.* at 776.
34 *Id.*
35 *Id.* at 771.
36 *Id.* at 774.
37 *Id.* at 776.
38 *Id.* at 776.
39 *Loveland v. Orem City Corp.*, 746 P.2d 763 (Utah 1987).
40 *Id.* at 776 (holding the City's activities to be within the exercise of a government function, thereby relieving the City from all liability from this matter).
41 *Id.* at 770.
42 *Id.* at 770.
43 *Jones v. Gillen*, 564 So. 2d 1274 (La. Ct. App.) *writ denied*, 568 So. 2d 1080 (La. 1990) and *writ denied*, 568 So. 2d 1081 (La. 1990).

44 *Id.* at 1276.
45 *Id.*
46 *Id.* at 1277.
47 *Id.*
48 *Id.*
49 *Id.*
50 *Id.* at 1282.
51 *Id.*
52 *Id.*
53 Dobbs, DB, Hayden, PT, Bublick EM. 2011. *The Law of Torts* (2nd ed.), West, St. Paul, MN, § 124.
54 *Id.* at § 125.
55 *Id.* at § 285.
56 *Id.* at § 148.
57 *Id.* at § 273.
58 *Id.* at § 161.
59 *Id.* at § 127.
60 *Id.*
61 *Id.* at § 183.
62 *Id.* at § 186.
63 *Id.*
64 *Id.*
65 *Id.* at § 198.
66 *Id.* at § 205.
67 *Id.* at § 208.
68 *Id.* at § 726.
69 *Id.*
70 Id. (Finding that courts may sometimes conclude that the plaintiff's conduct is the sole proximate cause of her own harm, in which case recovery is altogether barred.)
71 *Id.*
72 *Id.* at § 220.
73 *Id.* at § 238.
74 *Id.* at § 446.
75 *Rickwalt v. Richfield Lakes Corp.*, 246 Mich. App. 450, 633 N.W.2d 418 (2001).
76 See *Garner v. City of New York*, 6 A.D.3d 387, 775 N.Y.S.2d 335 (2004) (holding that the private landowners of a private beach could not be held liable for the plaintiff's injuries without evidence that the landowners controlled or maintained the premises); *Adika v. Beekman Towers, Inc.*, 633 So. 2d 1170 (Fla. 3rd DCA 1994); *Sperka v. Little Sabine Bay, Inc.*, 642 So. 2d 654 (Fla. 1st DCA 1994); *Darden v. Pebble Beach Realty, Inc.*, 860 F. Supp. 1101 (E.D.N.C. 1993)
77 Rickwalt, 246 Mich. App. at 453.
78 *Id.*
79 *Id.*
80 The Court of Appeals of Michigan also reversed in part the trial court's awards of costs and interest. *Id.*
81 See *Gonzales v. City of San Diego*, 130 Cal. App. 3d 882, 182 Cal. Rptr. 73 (Ct. App. 1982) (holding that the city was not immune from liability under the applicable sovereign immunity statute); *Casoni v. Town of Islip*, 278 A.D. 715, 103 N.Y.S.2d 435 (App. Div. 1951); *Piggott v. United States*, 480 F.2d 138 (4th Cir. 1973); *Corbett v. City of Myrtle Beach, S.C.*, 336 S.C. 601, 521 S.E.2d 276 (Ct. App. 1999); *Breaux v. City of Miami Beach*, 899 So. 2d 1059 (Fla. 2005); *Andrews v. Dep't of Natural Res., State of Fla.*, 557 So. 2d 85 (Fla. 2nd DCA 1990); *Butler v. Sarasota Cnty.*, 501 So. 2d 579 (Fla. 1986); *Hall v. Lemieux*, 378 So. 2d 130 (La. Ct. App. 1979) *writ denied*, 381 So. 2d 1220 (La. 1980)

An Overview of Aquatic Law 19

82 *Metro. Dade Cnty. v. Zapata*, 601 So. 2d 239, 241 (Fla. 3rd DCA 1992).
83 *Id.*
84 *Id.*
85 *Id.*
86 However, the Third District Court of Appeal reversed the lower decision, and remanded to the trial court for a new trial because of improper final argument made by plaintiff's counsel. *Id.* (Metro at 241.)
87 See *Poleyeff v. Seville Beach Hotel Corp.*, 782 So. 2d 422 (Fla. 3rd DCA 2001) (holding that the defendant hotel did not owe a duty to protect the swimmer from dangerous rip tides in the ocean); *Brown v. City of Vero Beach*, 64 So. 3d 172 (Fla. 4th DCA 2011) (holding that the city and county were immune from liability for injury or death caused by the naturally occurring conditions along the coastline); *Cutler v. City of Jacksonville Beach*, 489 So. 2d 126 (Fla. Dist. Ct. App. 1986); *Scott v. City of Long Beach*, 109 Cal. App. 254, 292 P. 664 (Cal. Ct. App. 1930); *Stann v. Waukesha Cnty.*, 161 Wis. 2d 808, 468 N.W.2d 775 (Ct. App. 1991); *Covington v. United States*, 902 F. Supp. 1207 (D. Haw. 1995); *Dewick v. Vill. of Penn Yan*, 972 F. Supp. 166 (W.D.N.Y. 1997); *Pelz v. City of Clearwater*, 568 So. 2d 949 (Fla. 2nd DCA 1990); *Warren v. Palm Beach Cnty.*, 528 So. 2d 413 (Fla. 4th DCA) *cause dismissed*, 537 So. 2d 570 (Fla. 1988); *Vela v. Cameron Cnty.*, 703 S.W.2d 721 (Tex. App. 1985), *writ refused NRE* (June 4, 1986).
88 *Suarez v. City of Texas*, No. 13-0947, 2015 WL 3802865 (Tex. June 19, 2015).
89 *Id.* at *3.
90 *Id.* at *3.
91 *Id.* at *3.
92 *Id.* at *3.
93 See Tex. Civ. Prac. & Rem. Code Ann. § 75.002 (West).
94 N.H. Rev. Stat. Ann. § 508:14(I).
95 *Coan v. New Hampshire Dep't of Envtl. Servs.*, 161 N.H. 1, 8 A.3d 109 (2010).
96 *Id.* at 8.
97 *Id.* at 3.
98 See *Kenison v. Dubois*, 152 N.H. 448, 879 A.2d 1161 (2005).
99 See Haw. Rev. Stat. § 520-1 (West); Ariz. Rev. Stat. Ann. § 33-1551; Tex. Civ. Prac. & Rem. Code Ann. § 75.002 (West); Wash. Rev. Code Ann. § 4.24.210 (West); La. Rev. Stat. Ann. 9:2795; N.D. Cent. Code Ann. § 53-08-01 (West); Ind. Code Ann. § 14-22-10-2 (West); Fla. Stat. Ann. § 375.251 (West); Mo. Ann. Stat. § 537.346 (West).
100 Cal. Civ. Code § 846 (West).
101 *Id.*
102 See *Stone Mountain Mem'l Ass'n v. Herrington*, 225 Ga. 746, 171 S.E.2d 521 (1969) (holding that a $2 parking fee was not considered a charge under the Georgia recreational use statute); *Majeske v. Jekyll Island State Park Auth.*, 209 Ga. App. 118, 433 S.E.2d 304 (1993).
103 *Cole v. S. Carolina Elec. & Gas, Inc.*, 355 S.C. 183, 584 S.E.2d 405 (Ct. App. 2003) *aff'd as modified sub nom. Cole v. S. Carolina Elec. & Gas, Inc.*, 362 S.C. 445, 608 S.E.2d 859 (2005).
104 *Id.*
105 *Id.* at 188.
106 *Id.*
107 *Id.*
108 *Rivera v. Philadelphia Theological Seminary of St. Charles Borromeo, Inc.*, 510 Pa. 1, 17, 507 A.2d 1, 9 (1986).
109 *Id.*
110 See *Williams v. City of Baton Rouge*, 200 So. 2d 420 (La. Ct. App.) *writ issued*, 251 La. 43, 202 So. 2d 656 (1967) and *writ issued*, 251 La. 45, 202 So. 2d 656 (1967) and *aff'd and amended*, 252 La. 770, 214 So. 2d 138 (1968); *City of Evansville v. Blue*, 212 Ind. 130, 8 N.E.2d 224 (1937); *Estate of A.R. v. Grier*, No. CIV.A. H-10-0533, 2011 WL 3813253 (S.D. Tex. Aug. 26, 2011), *aff'd*

(Oct. 16, 2013); *Williams v. Chicago Bd. of Educ.*, 267 Ill. App. 3d 446, 642 N.E.2d 764 (1994); *Estate of C.A. v. Grier*, No. CIV.A. H-10-00531, 2011 WL 3902750, at *1 (S.D. Tex. Sept. 6, 2011) *aff'd sub nom. Estate of C.A. v. Castro*, 547 F. App'x 621 (5th Cir. 2013).
111 *Dontas v. City of New York*, 183 A.D.2d 868, 584 N.Y.S.2d 134 (1992).
112 *Id.* at 869.
113 *Id.*
114 *Id.*
115 However, on appeal the $2 million in damages for conscious pain and suffering of the drowning victim was found excessive. Also the $4 million in damages for wrongful death was excessive, resulting in the jury verdict being reversed and a new trial granted on damages. *Id.*
116 *Ward v. City of Millen*, 162 Ga. App. 148, 290 S.E.2d 342 (1982).
117 *Id.*
118 *Ward*, 162 Ga. App. at 149.
119 *Id.*
120 *Id.*
121 *Id.*
122 See *Young Men's Christian Ass'n of Metro. Atlanta v. Bailey*, 107 Ga. App. 417, 130 S.E.2d 242 (1963); *Thompson v. Ewin*, 457 So. 2d 303 (La. Ct. App.) *writ denied*, 460 So. 2d 1043 (La. 1984); *Nelson v. Thibaut HG Corp.*, 2007-0515 (La. App. 4 Cir. 1/23/08), 977 So. 2d 1055 (La. Ct. App. 2008); *Barnett v. Ludwig & Co.*, 2011 IL App (2d) 101053, 960 N.E.2d 722; *Hecht v. Des Moines Playground & Recreation Ass'n*, 227 Iowa 81, 287 N.W. 259 (1939); *Adler v. Copeland*, 105 So. 2d 594 (Fla. 3rd DCA 1958); *Banks v. Mason*, 132 So. 2d 219 (Fla. 2nd DCA 1961); *Avery v. Morse*, 149 Misc. 318, 267 N.Y.S. 210 (Sup. Ct. 1933); *Mullens v. Binsky*, 130 Ohio App. 3d 64, 719 N.E.2d 599 (1998); *Simmons v. Whittington*, 444 So. 2d 1357 (La. Ct. App.) *writ denied*, 447 So. 2d 1071 (La. 1984); *Hemphill v. Johnson*, 230 Ga. App. 478, 497 S.E.2d 16 (1998); *Spergel v. Dolphin Pools Corp.*, 741 F. Supp. 8 (D.D.C. 1990); *St. Hill v. Tabor*, 542 So. 2d 499 (La. 1989); *Moses v. Bridgeman*, 355 Ark. 460, 139 S.W.3d 503 (2003); *Smith v. Jung*, 241 So. 2d 874 (Fla. Dist. Ct. App. 1970); *Bell v. Page*, 2 N.C. App. 132, 162 S.E.2d 693 (1968); *Hemispheres Condo. Ass'n, Inc. v. Corbin*, 357 So. 2d 1074 (Fla. 3rd DCA 1978).
123 *Coleman v. Shaw*, 281 S.C. 107, 314 S.E.2d 154 (Ct. App. 1984).
124 See South Carolina Department of Health and Environmental Control Regulation 61-51(B)(10) (stating that no swimming pool shall allow solo bathing and at least one employee, having first-aid training, shall be monitoring the pool at all times).
125 *Coleman*, 281 S.C. at 109.
126 *Id.*
127 *Id.*
128 *Id.* at 113.
129 *Id.* at 110.
130 *Id.*
131 *Odom v. Lee*, 145 Ga. App. 304, 243 S.E.2d 699 (1978).
132 *Id.*
133 *Id.*
134 *Id.*
135 *Id.*
136 24 No. 1 Verdicts, Settlements & Tactics art. 33 (2004).
137 *Id.*
138 *Id.*
139 *Id.*
140 *Id.*
141 *Id.*
142 *Id.*

An Overview of Aquatic Law 21

143 See *Allen v. William P. McDonald Corp.*, 42 So. 2d 706 (Fla. 1949); *Cole v. S. Carolina Elec. & Gas, Inc.*, 355 S.C. 183, 584 S.E.2d 405 (Ct. App. 2003) *aff'd as modified sub nom. Cole v. S. Carolina Elec. & Gas, Inc.*, 362 S.C. 445, 608 S.E.2d 859 (2005); *Gilbertson v. Lennar Homes, Inc.*, 629 So. 2d 1029 (Fla. Dist. Ct. App. 1993).
144 *Simons v. Hedges*, 286 A.D. 1044, 144 N.Y.S.2d 828 (App. Div. 1955).
145 *Id.* at 1045.
146 *Id.*
147 *Id.*
148 *Id.*
149 *Id.*
150 *Leonard v. Peoples Camp Corp.*, 9 A.D.2d 420, 194 N.Y.S.2d 863 (1959) *aff'd*, 9 N.Y.2d 652, 173 N.E.2d 46 (1961).
151 *Id.*
152 *Id.* at 421.
153 *Id.* at 422.
154 *Id.*
155 *Id.*
156 See *Torf v. Commonwealth Edison*, 268 Ill. App. 3d 87, 644 N.E.2d 467 (1994) (holding that the water discharged into the lake was an open and obvious risk, preventing the city from being held liable for the drowning death of a minor); *Carlson v. City of Pascagoula*, 227 So. 2d 279 (Miss. 1969) (holding that the city was not negligent in failing to provide a lifeguard); *Saga Bay Prop. Owners Ass'n v. Askew*, 513 So. 2d 691 (Fla. 3rd DCA 1987); *Kinya v. Lifter, Inc.*, 489 So. 2d 92 (Fla. 3rd DCA 1986); *Longmore v. Saga Bay Prop. Owners Ass'n, Inc.*, 868 So. 2d 1268 (Fla. 3rd DCA 2004); *Brazier v. Phoenix Grp. Mgmt.*, 280 Ga. App. 67, 633 S.E.2d 354 (2006); *Cortes v. State*, 191 Neb. 795, 218 N.W.2d 214 (1974); *Clem v. United States*, 601 F. Supp. 835 (N.D. Ind. 1985); *Navarro v. Country Vill. Homeowners' Ass'n*, 654 So. 2d 167 (Fla. 3rd DCA 1995) (holding that the deep-water drop off in the lake did not constitute a dangerous condition under the law); *Jones v. Interstate N. Associates*, 145 Ga. App. 366, 243 S.E.2d 737 (1978); *Plummer v. Bd. of Comm'rs of St. Joseph Cnty.*, 653 N.E.2d 519 (Ind. Ct. App. 1995).
157 *Blacka v. James*, 205 Va. 646, 139 S.E.2d 47 (1964).
158 *Id.* at 647.
159 *Id.* at 648.
160 *Id.*
161 *Id.* at 651.
162 *Hutchinson v. Twp. of Portage*, No. 240136, 2003 WL 21958278, (Mich. Ct. App. Aug. 14, 2003).
163 *Id.* at *1.
164 *Id.*
165 *Id.*; see M.C.L. § 691.1407(2).
166 *Hutchinson*, at *1.
167 *Trinity River Auth. v. Williams*, 689 S.W.2d 883 (Tex. 1985).
168 *Id.* at 884.
169 *Id.*
170 *Id.*
171 *Id.*
172 *Id.* at 885.
173 *Id.*
174 *Id.*
175 *Id.* at 886.
176 *Id.*
177 See *Hendershot v. Kapok Tree Inn, Inc.*, 203 So. 2d 628 (Fla. 2nd DCA 1967); *Smith v. U. S. Steel Corp.*, 351 So. 2d 1369 (Ala. 1977) (holding that the landowner was not liable for the drowning deaths of the two children because they were trespassers); *Kaweblum ex rel. Kaweblum v.*

Thornhill Estates Homeowners Ass'n, Inc., 801 So. 2d 1015 (Fla. 4th DCA 2001); *Newby v. W. Palm Beach Water Co.,* 47 So. 2d 527 (Fla. 1950); *Howard v. Atl. Coast Line R. Co.,* 231 F.2d 592 (5th Cir. 1956); *St. Clair v. City of Macon,* 43 Ga. App. 598, 159 S.E. 758, 758 (1931); *Jones v. Gillen,* 504 So. 2d 575 (La. Ct. App.) *writ denied,* 508 So. 2d 86 (La. 1987); *Hammond v. Realty Leasing, Inc.,* 342 So. 2d 915 (Ala. 1977); *Prince v. Wolf,* 93 Ill. App. 3d 505, 417 N.E.2d 679 (1981); *Lerma v. Rockford Blacktop Const. Co.,* 247 Ill. App. 3d 567, 617 N.E.2d 531 (1993); *Ochampaugh v. City of Seattle,* 91 Wash. 2d 514, 588 P.2d 1351 (1979); *Fickling v. City Council of Augusta,* 110 Ga. App. 330, 138 S.E.2d 437 (1964); *Eades v. Am. Cast-Iron Pipe Co.,* 208 Ala. 556, 94 So. 593 (1922); *McLean v. Ward,* 1 N.C. App. 572, 162 S.E.2d 64 (1968); *Hill v. Guy,* 161 Mich. App. 519, 411 N.W.2d 757 (1987).
178 *White v. Georgia Power Co.,* 265 Ga. App. 664, 595 S.E.2d 353 (2004).
179 *Id.*
180 *Id.* at 667.
181 *Id.*
182 *Id.*
183 *Webb v. Hunter,* 431 So. 2d 1131 (Miss. 1983).
184 *Id.*
185 *Id.* at 1132.
186 *Id.*
187 *Id.*
188 *Id.* at 1133.
189 *McFarland v. Grau,* 305 S.W.2d 91 (Mo. Ct. App. 1957).
190 *Id.* at 94.
191 *Id.*
192 *Id.* at 95.
193 *Id.*
194 *Id.*
195 *Id.* at 96.
196 *Id.* at 98.
197 *Id.*
198 *Id.* at 101.
199 *Id.*
200 *Id.*
201 *Id.*
202 *Garber v. Prudential Ins. Co. of Am.,* 203 Cal. App. 2d 693, 22 Cal. Rptr. 123 (Ct. App. 1962).
203 *Id.*
204 *Id.*
205 *Id.* at 131.
206 *Id.*
207 *Metro. Dade Cnty. v. Zapata,* 601 So. 2d 239, 241 (Fla. 3rd DCA 1992).
208 *Id.*
209 *Id.* at 242; See *Blackburn v. Dorta,* 348 So.2d 287 (Fla. 1977); *Mazzeo v. City of Sebastian,* 550 So.2d 1113 (Fla. 1989).
210 *Carter v. Boys' Club of Greater Kansas City,* 552 S.W.2d 327, 332 (Mo. Ct. App. 1977).
211 *Id.* at 329.
212 *Id.*
213 *Id.* at 333.
214 *Leakas v. Columbia Country Club,* 831 F. Supp. 1231 (D. Md. 1993).
215 *Id.* at 1241.
216 *Id.* at 1238.
217 *Id.*
218 *Kalm, Inc. v. Hawley,* 406 S.W.2d 394 (Ky. 1966).

219 *Id.* at 395 (Kalm).
220 *Id.*
221 *Id.*
222 *Id.*
223 *Id.*
224 *Id.*
225 *Saga Bay Prop. Owners Ass'n v. Askew*, 513 So. 2d 691 (Fla. Dist. Ct. App. 1987).
226 *Id.* at 693.
227 *Id.* at 692.
228 *Id.* at 693.
229 *Id.*
230 *Smith v. Chicago & E. I. R. Co.*, 342 Ill. App. 78, 95 N.E.2d 95 (Ill. App. Ct. 1950).
231 *Id.* at 82.
232 *Id.* at 86.
233 *Woolridge v. E. Texas Baptist Univ.*, 154 S.W.3d 257, 259 (Tex. App. 2005).
234 See *Hendershot v. Kapok Tree Inn, Inc.*, 203 So. 2d 628 (Fla. Dist. Ct. App. 1967).
235 "Courts have discussed whether, and under what circumstances, aviation exclusion clauses in policies of life or accident insurance apply where the insured was injured or died from injuries or conditions encountered after he left an aircraft upon termination of a successful flight, or after he escaped without serious injury from a disabled aircraft either by ejecting himself from the plane during a flight or by escaping from the plane after it crash-landed." *Id.*
236 See *Neel v. Mutual Life Ins. Co.*, 131 F.2d 159 (2d Cir. 1942); *Green v. Mutual Benefit Life Ins. Co.*, 144 F.2d 55 (1st Cir. 1944); *Hobbs v. Franklin Life Ins. Co.*, 253 F.2d 591 (5th Cir. 1958); *McDaniel v. Standard Acc. Ins. Co.*, 221 F.2d 171 (7th Cir. 1955).
237 *Howard v. Equitable Life Assur. Soc. of U. S.*, 360 Mass. 424, 274 N.E.2d 819, 820 (1971).
238 See *Howe v. Prudential Ins. Co. of Am.*, 130 Ga. App. 107, 202 S.E.2d 669 (1973) rev'd, 232 Ga. 1, 205 S.E.2d 263 (1974); *Elliott v. Massachusetts Mut. Life Ins. Co.*, 388 F.2d 362 (5th Cir. 1968); *Rossman v. Metro. Life Ins. Co.*, 71 F. Supp. 592, 597 (D. Me. 1947) (holding that the plaintiff was only entitled to the reserve value of the insurance policy); *Barringer v. Prudential Ins. Co. of Am.*, 62 F. Supp. 286 (E.D. Pa. 1945) aff'd, 153 F.2d 224 (3d Cir. 1946); *Order of United Commercial Travelers of Am. v. King*, 161 F.2d 108 (4th Cir. 1947) aff'd, 333 U.S. 153, 68 S. Ct. 488, 92 L. Ed. 608 (U.S.S.C. 1948); *Rauch v. Underwriters at Lloyd's of London*, 320 F.2d 525 (9th Cir. 1963); *Goforth v. Franklin Life Ins. Co.*, 202 Kan. 413, 449 P.2d 477 (1969); *Wendorff v. Missouri State Life Ins. Co.*, 318 Mo. 363, 1 S.W.2d 99 (1927).
239 *Howard*, 360 Mass. at 425.
240 *Id.*
241 *Id.*
242 *Id.* at 426
243 See *McDaniel v. Standard Acc. Ins. Co.*, 221 F.2d 171 (7th Cir. 1955) (holding that the aviation exclusion clause did not apply because the plaintiff escaped the airplane and swam a considerable distance in cold water before drowning); *Sec. Mut. Life Ins. Co. v. Hollingsworth*, 1969 OK 126, 459 P.2d 592 (stating that an essential fact in reaching its conclusion was that upon the flight terminating, the passengers were uninjured and in a position of at least potential safety).
244 See *Spady v. Bethlehem Area Sch. Dist.*, No. CIV.A. 12-6731, 2014 WL 3746535, at *7 (E.D. Pa. July 30, 2014).
245 *Corda v. Brook Valley Enterprises, Inc.*, 63 N.C. App. 653, 306 S.E.2d 173 (1983).
246 *Id.* at 655.
247 *Id.*
248 *Id.* at 656.
249 *Id.*
250 *Id.*
251 *Id.*

252 *Spady v. Bethlehem Area Sch. Dist.*, No. CIV.A. 12-6731, 2014 WL 3746535 (E.D. Pa. July 30, 2014).
253 *Id.* at *1.
254 *Id.*
255 *Id.*
256 *Id.*
257 *Id.* at *2.
258 *Id.*
259 *Id.* at *4.
260 *Id.* at *6.
261 *Id.* at *7.

2 Legal Concepts, Drowning, and Lifeguard Effectiveness

John R. Fletemeyer

CONTENTS

Legal Concepts 26
    Torts 26
        Contributory Negligence 26
        Comparative Negligence 27
        Elements of Tort Law 27
Drowning 28
    The Cost of Drowning 29
    Clinical Death vs. Biological Death 30
    Distress 30
    Salt vs. Fresh Water Drowning 32
    Dry Drownings 33
    Drowning Physiology 33
        Stage 1, Initial Apnea 34
        Stage 2, Dyspnea 34
        Stage 3, Terminal Apnea 34
        Stage 4, Cardiac Arrest 35
    Passive/Silent Drowning 38
    Cold Water Drowning 39
    Shallow Water Blackout 39
    Toddler Drowning 40
    Entrapment Drownings 41
    Construction and Work-Related Drowning 41
    SCUBA and Snorkeling Drowning 41
Police and First Responders 42
Lifeguards 43
    YMCA 46
    American Lifeguard Association 46
    American Red Cross 47
    United States Lifesaving Association (USLA) 47
References 47

The objective of this chapter is to provide attorneys and aquatic professionals information about the following:

- Legal concepts germane to aquatic law
- The physiology of drowning
- The behavioral correlates transpiring during a drowning event
- Estimating the cost of fatal and non-fatal drowning involving negligence
- The effectiveness of various drowning prevention methods, including lifeguard practices

LEGAL CONCEPTS

Much of what is presented in the beginning of this chapter will be familiar to practicing attorneys. However, since this book is also intended for other professionals, including aquatic supervisors, administrators, risk managers, and insurance underwriters, the following includes some important legal concepts and definitions related to allegations of a breach of duty when a serious aquatic accident or drowning is involved.

TORTS

The American judicial system provides members of the public with the opportunity to be compensated following an injury or death where negligence can be proven to the courts. In the majority of cases, the compensation is money. In most aquatic cases, death or injury resulting in lifelong care is often involved. Consequently, compensation is often considerable.

Most aquatic injury and fatal drowning cases seldom go to trial. The risk of a huge jury award is too great. Instead, statistics reveal that a majority of negligence claims are settled by negotiated settlement that often relies on opinions from a host of experts.

The court defines negligence as conduct that falls below the standard of care that society expects of a reasonably prudent person acting under similar circumstances. If a person sustains an injury or a family member drowns, and if the injury or drowning was the result of unreasonable conduct, then the party making the claim is entitled to compensation. Since the vast majority, perhaps 90 percent or greater, are considered preventable, claims of negligence are to be expected.

In many cases, a claim of "gross negligence" is involved. If proven, this elevates the opportunity for financial recovery and in some cases and some states, may also include additional awards taking the form of "punitive damages." *Gross negligence* is defined as a conscious and voluntary disregard of the need to use reasonable care, which is likely to cause foreseeable grave injury or harm to an individual.

Contributory Negligence

Contributory negligence places the blame for an accident or fatal drowning on more than one cause. The concept of contributory negligence allows for the defense to argue that the injured plaintiff caused or at least contributed to his or her accident or drowning. Although some states still accept this as a defense argument, most states have replaced it with "comparative" negligence.

Comparative Negligence

Comparative negligence places fault on a percentage basis. Consequently, if the courts determine that the actions of the injured were 25 percent responsible for an accident, the injured victim is entitled to 75 percent of the jury award. In some cases there may be several comparative factors. For example, in a springboard diving case, the injured victim and his actions might be placed at 20 percent fault, the coach at 40 percent, the lifeguards at 20 percent, and the pool contractor at 20 percent.

Elements of Tort Law

In regards to tort law, there are four elements that require consideration, including duty, breach of duty, proximate cause or causes, and damages.

Duty

Duty requires that the plaintiff establish that the defendant owed the plaintiff (the injured) a duty or standard of care. Duty is a legal obligation to act in accord with a standard of conduct designed to protect the plaintiff from an unreasonable risk or hazard.

With regards to aquatic injury and fatal drowning cases, defining duty is sometimes challenging, even for an experienced attorney. Duty is often defined differently from state to state and sometimes is predicated on a consensus opinion by aquatic professionals. Only in rare cases are there national laws that clearly define duty; for example, the Virginia Grahame Baker Act requires swimming pools to be equipped with anti-entrapment drains.

Breach of Duty

Breach of duty requires that the plaintiff prove that the defendant did something a reasonable person owning the duty would not do, or didn't do something that the same reasonable person would not do.

In regards to claims of aquatic breach of duty, several examples come to mind. A breach of duty is failing to warn the public about a life-threatening hazard. Often this involves failing to provide strategically located warning signs. It may involve the failure to maintain swimming pool water chemistry at a certain level, allowing the water to become cloudy. It may involve allowing pool or beach guests to consume alcohol on the premises. It may involve a lifeguard engaging in distractive behavior, preventing him or her from watching bathers in his or her assigned area.

Proximate Cause

The proximate cause is the primary cause of the injury or accident. Proximate cause produces a particularly foreseeable consequence without the intervention of any independent or foreseeable cause or causes. Some states recognize that there may be more than one proximate cause.

The following represents examples of a proximate cause where courts found the defendant negligent:

- A lifeguard reading a newspaper and failing to observe a victim that was actively drowning.
- A diving coach failing to follow concussion protocol when a diver is suspected of sustaining a head injury.
- Lifeguards failing to clear the water during an electrical storm.
- A hotel pool failing to repair a self-latching gate, allowing a toddler to access the pool and then drowning.

Damages

If a plaintiff has established that a duty has been breached, the plaintiff must demonstrate that he or she has suffered damages. In the case of a fatal drowning event, this might also include family members suffering both an emotional loss and an economic loss. In the case of pain and suffering, this is often subjective, with courts often granting significant latitude when making this claim. If a father drowned due to negligence, there is likely an economic loss (the loss income) if proven that he was the breadwinner. In the case of a spinal injury, the loss may be long-term, with the victim requiring lifelong medical care. In the chapter to follow, in addition, there might also be a need for long-term psychological intervention that may or may not be recognized as a legitimate claim, depending on the state and court.

DROWNING

Fatal drowning involves several complex and interactive physiological responses leading to clinical and eventually biological death. Attempting to explain to a jury how a person drowns and the economic costs associated with a drowning is challenging even for the most experienced attorney. In most cases, this can only be accomplished with the help of an expert.

The definition of drowning adopted by World Congress on Drowning in 2002 and accepted by the World Health Organization in 2005, and then later by the CDC, the American Red Cross, and the YMCA of the USA, states that "drowning is a process of experiencing respiratory impairment from submersion/immersion in liquid." Accepting this definition requires using the term "fatal drowning" if a victim dies from a drowning event. In addition, drowning outcomes are classified as either "death," "morbidity," or "no morbidity."

Drowning is one of the leading causes of accidental death. Some of the major causes of drowning include:

- Failure to recognize and respond to a hazard.
- Failure to recognize drowning behavior and perform timely rescue of a drowning victim.
- Lack of responsible adult supervision of a toddler or a young child.
- Inability to escape from a dangerous situation.

Legal Concepts, Drowning, and Lifeguard Effectiveness

- Failing to provide proper supervision involving group activities and individual supervision among toddlers.
- Lack of knowledge, skills, or ability necessary to rescue a drowning victim.
- Not knowing CPR or how to use an automated external defibrillator (AED).

Drownings occur in all aquatic environments but are most common is swimming pools and among children. Some unusual examples include bathtubs, toilets, storm drains, water buckets, and wells. Figure 2.1 uses CDC statistics to create a pie graph demonstrating where drownings are likely to occur.

THE COST OF DROWNING

The cost of a drowning, whether fatal or non-fatal, is difficult to calculate. This is especially the case when a non-fatal drowning is responsible for long-term care resulting from brain damage or a spinal injury. When an individual dies as a result of the negligence, damages can often be recovered as a result of legal action. Economic recovery often varies significantly from state to state. Some states have imposed statutory caps on the amount that can be awarded. In addition, in some states certain entities are protected from liability by sovereign immunity. Generally speaking, if negligence can be proven, the following damages may be recovered.

- Economic damages. These damages are documentable financial losses incurred by the real parties in interest. They may include medical bills incurred before the death of the decedent, expected loss earnings of the decedent over their lifetime, lost benefits such as pensions and even healthcare, and lost value of services and goods a decedent would have provided otherwise had he or she not died. The amount of recoverable based on these types of economic damages will vary widely from case to case based on

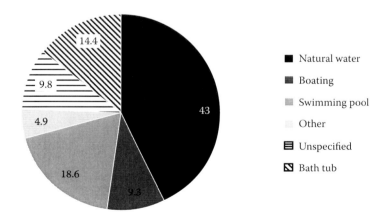

FIGURE 2.1 CDC statitics showing aquatic environment and the relative percentage of fatal drownings occuring in each category.

documentable factors present at the time of death of the decedent and possibly based on calculations for future loses.
- Non-economic damages. These damages, which are non-pecuniary in nature but are nonetheless losses sustained by real parties of interest, can include pain, suffering, mental anguish, and psychological trauma sustained by survivors. Additionally, survivors can recover compensation for loss of consortium, loss of companionship, and a host of other intangible losses and harms sustained by survivors.
- Punitive damages. Another element of compensation may be what is known as punitive damages, which may potentially be assessed in cases involving intentional or egregious negligence causing death. Other compensation recoverable may include interest and legal fees.

Estimating the total cost of a non-fatal drowning is even more difficult to determine and is often the subject of debate and focus of mediation. As mentioned in the introduction and in Chapter 12, the psychological cost of drowning must be considered, as this may represent years or even a lifetime of costs when a spinal injury is involved or when the victim suffers permanent brain damage.

CLINICAL DEATH VS. BIOLOGICAL DEATH

In cases involving resuscitation efforts, it is important to understand the difference between clinical and biological death. Clinical death begins once a victim submerges underwater and stops breathing. Shortly thereafter, the heart usually begins to fibrillate and then stops. While this is occurring, it is possible to successfully resuscitate a victim by administering CPR and using an AED. The timeline for this varies and is usually estimated between four and six minutes. However, the timeline can be significantly extended but not indefinitely if the drowning event occurs in cold water, especially among children. Sometime between six and ten minutes is the onset of biological death. This is when there is irreparable and irreversible brain damage and when emergency first aid intervention is no longer possible.

DISTRESS

Experts providing court testimony often differentiate between distress and drowning and in addition, characterize drowning as being either passive or active.

Distress involves a victim in trouble, but who still has the ability to keep afloat. A distressed victim is usually able to signal or yell for help and take actions directed toward self-rescue. As a distressed victim begins to hyperventilate, he or she devotes an exerted effort to stay afloat before progressing to the active drowning stage.

Drowning usually but not always follows the initial stage of distress as described previously. Unless a victim suddenly becomes unconscious, which seldom happens, drowning progresses through four stages and sometimes lasts several minutes. These stages are discussed in detail below. Among children, the timeline often becomes abbreviated but usually not significantly. Drowning falls into two categories and it is critical for an expert to determine which of the two is involved.

FIGURE 2.2 A nine-year-old begins to "actively" drown.

Active drowning. Active drowning usually involves a non-swimmer, but not always. In an active drowning, the victim becomes exhausted, or if occurring in cold water, "hypothermic." Before submerging, the victim struggles—but not how it is portrayed by the movie and TV industries. A victim seldom if ever yells for help and never begins to wildly panic. Active drowning behavior among children and adults is usually different. Among children, actively drowning is "quiet" and occurs quickly with few if any overt signs of distress or panic (see Figure 2.2). Among adults, active drowning may last several minutes, accompanied by several behaviors related to panic but not involving the wild flailing of arms and legs, splashing, and yelling for help (see Figure 2.3).

Some "typical" behaviors involved in an active drowning include the following:

- Victim's head becomes lower and lower in the water. Victim may be hyperventilating and gasping for air as a result from undergoing an initial stage of distress or, in some cases, may be attempting to hold their breath.
- Victim's eyes are usually fixated toward the shore or the swimming pool deck. Eyes are often described as being "glassy."
- Victim is usually vertical in the water with his or her arms moving up and down. Sometimes this behavior is called "climbing the ladder."
- In some cases, the victim will attempt to roll over on his or her back but usually without success.
- Victim usually attempts to move toward the shore but without any forward progress.
- If the victim has long hair, it is usually covering his or her eyes and forehead.
- Among young children, drowning sometimes suggests play behavior. This may cause an adult or lifeguard to fail in the recognition that a child is drowning and not playing.
- In some active drowning events, a victim will surface, submerge, and surface several times before finally submerging underwater for the final time.

FIGURE 2.3 An adult begins to "actively" drown.

Note that most drowning victims are initially negatively buoyant and sink to the pool or lake bottom. Only when decomposition begins days later, causing gases to form in the tissues, will a body become positively buoyant and float to the surface. In rare cases bottom suction will "hold" the body on the bottom. There have been several documented reports of when thunderstorms have released a body from the bottom. A few times dynamite has been successful in forcing a body to float to the surface.

Salt vs. Fresh Water Drowning

Understanding drowning is further complicated by whether the drowning event occurs in fresh or salt water. If occurring in fresh water, when water is taken into the lungs it will be pulled into the pulmonary circulation by osmosis. The dilution of blood levels leads to hemolysis (or the bursting of red blood cells). The resulting elevation of the plasma K+ (potassium) level and depression of the Na+ (sodium) level alter the electrical activity of the heart, often causing ventricular fibrillation. In animal experiments, this effect was shown to be capable of causing cardiac arrest in two to three minutes. Note that the reliability using animal drowning experiments, primarily dogs, rats, and primates, to understand drowning physiology among humans is suspect because the physiology of these animals is different from human physiology.

Drowning in cold water often dramatically changes survival outcomes, especially among children. In cold water, survival times are usually significantly increased when the mammalian diving reflex occurs. This reflex protects the body by putting it into an energy saving mode to maximize the time it can stay underwater. This reflex involves blood being shunted away from the extremities and being focused to the vital organs, thus delaying and prolonging life.

When a drowning occurs in sea water, the opposite physiological response occurs. Sea water is hypertonic to blood (more salty). Osmosis will instead pull water from the bloodstream into the lungs, thickening the blood. Results from animal experiments finds that the thicker blood requires more work from the heart, leading to cardiac arrest in eight to ten minutes.

Secondary drowning represents another category of death. A consensus among experts is that this category should not be added to the annual count of fatal drownings that occur in the United States each year. For this reason, and for several others that are discussed in this book, the number of drownings that occur in the United States may be significantly underestimated (refer to the introduction of this book).

In a secondary drowning, inhaled fluid acts as an irritant inside the lungs. The physiological response to even small quantities includes the extrusion of liquids into the lungs. The medical term for this is *pulmonary edema*. When water enters the lungs, this reduces the ability to exchange air and leads to a drowning in the body's own fluids.

DRY DROWNINGS

In theory, a dry drowning results when a victim's vocal cords spasm, preventing water from entering the lungs. In medical terminology this is called a *laryngospasm*. Dry drownings were identified after observing dogs that were anesthetized and intentionally drowned. Fletemeyer reported in 1999 that dry drownings are exceeding rare, accounting for less than 10 percent of fatal drownings. More recent research rejects the existence of dry drowning, although theoretically it is possible. With the absence of water in the lungs, it is possible that the victim suffered a predisposing event responsible for breathing to stop before the victim submerged and inhaled water, that is, cardiac arrest, electrocution, seizure, or some similar trauma.

DROWNING PHYSIOLOGY

The physiology of drowning is complicated and difficult to understand without the benefit of a medical background. This makes explaining drowning to a jury problematic because most jury members do not have a scientific background.

Nevertheless, it is critically important for attorneys involved in a drowning case to understand this process, and in some cases, it is important for a jury to understand this process at a basic level. Drowning physiology can be represented in four stages lasting several minutes until the onset of biological death (Fletemeyer, 1995). However, these stages are artificially derived and may not progress sequentially from one to the other. Before the onset of these four stages, the victim is struggling in the water and exhibiting some or all of the behaviors that have been identified previously.

Stage 1, Initial Apnea

This stage is characterized by a reflex reaction that causes the victim's glottis to close. This results in the stopping of breathing and is usually associated with a period of panic but not how it is commonly depicted in the movies or on TV. Physiological characteristics of this stage include the following:

- Blood pressure rises and adrenaline flow increases.
- Victim begins swallowing water, resulting in less buoyancy and a diminished capacity to breathe.
- As the victim become less buoyant his or her head can no longer be held above the water.
- With the resulting lack of breathing, the brain begins to experience degeneration because it cannot get sufficient oxygen to function properly (hypoxia).
- Acidosis begins due to the buildup of excess waste chemicals in the victim's blood stream.

Stage 2, Dyspnea

In this stage, the victim is still in a semi-conscious state, but begins to swallow more water. Usually water begins to enter the lungs, resulting in the deterioration of the alveoli and damage to the chemical known as surfactant. This chemical is responsible for reducing the surface tension on an aveola's membrane, allowing for easy exchange of oxygen and carbon dioxide. Note that unpublished research by Fletemeyer that involved interviewing victims who survived these two stages reports that victims experience excruciating pain that is described as unbearable accompanied by thoughts that their head is ready to explode. The following are characteristics of this stage:

- Aspiration of water into the lungs due to the failure of the swallow reflex.
- Vomiting.
- Frothing of the victim's mouth due to the mixture of surfactant and water in the lungs.
- Brain hypoxia continues to evolve and acidosis continues to cause a severe imbalance in the chemicals of the blood.

Stage 3, Terminal Apnea

As soon as the victim becomes unconscious and stops breathing, he or she has progressed to this stage. The following list represents characteristics of this stage.

- Brain hypoxia and acidosis continues.
- Tonic convulsions may occur in some cases, in which the victim's entire body becomes rigid, causing the victim to jerk involuntarily.

- The victim' sphincter muscle may relax, resulting in uncontrolled urination or defecation in some cases.

Stage 4, Cardiac Arrest

In this stage, the heart ceases to function and pump blood. Depending on the circumstances, the third and fourth stages might begin simultaneously; meaning that the heart and lungs (breathing) stop at the same time. However, as most trained first responders know, the heart may sometimes continue beating for some time, going into a state of "fibrillation" after breathing has stopped; in some cases as long as five minutes and in rare cases involving cold water, even longer. The causes of drowning extracted from 2006 data collected by the CDC include the following:

- 31 percent—Drowning and submersion while in natural water.
- 27.9 percent—Unspecified drowning.
- 14.5 percent—Drowning and submersion while in a swimming pool.
- 9.4 percent—Drowning and submersion while in a bathtub.
- 7.2 percent—Drowning and submersion following a fall into natural water.
- 6.3 percent—Other specified drowning and submersion.
- 2.9 percent—Drowning and submersion following a fall into a swimming pool.
- 0.9 percent—Drowning and submersion following a fall into a bathtub.

To reiterate what was discussed in the introduction of this book, there is good evidence that the number of drownings reported by the CDC is significantly underestimated and the belief that the number of drownings is declining may not be accurate. Regarding the number of near-drownings, estimating this number is even more problematic than estimating the number of fatal drownings but may be between 100,000 (+/− 25 percent) annually.

About 90 percent of drownings (fatal and non-fatal) occur in fresh water. Fatal drowning events are significantly more common among males, with males three to four times more likely to drown than females. The reason for this male bias is the tendency for males to be more boisterous and more prone to risk-taking behavior.

Among children one through twelve years old, drowning ranks as the second leading cause of accidental death. The number of drownings for 2012 were entered in Figure 2.4 to demonstrate the relative numbers for each groups. Following the age category of 1–4, categories 65–Above, 35–44, and 25–35 rank high while the category "Less than 1" is relatively low.

Besides describing drowning in four stages, drowning can be described as being either an active drowning or a passive drowning. It is critically important that attorneys understand the difference between the two because there are often different proximate causes involved.

Active drownings occur in every aquatic environment and they often demonstrate some notable differences when occurring in different environments. In a study involving more than 500 active drownings in the All-American Canal, Fletemeyer identified several contributing factors that were responsible for so many fatalities (see Table 2.1).

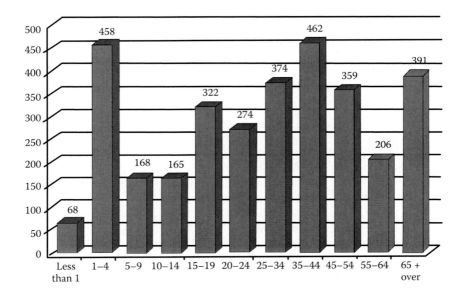

FIGURE 2.4 Number of drownings defined by age groups.

An active drowning usually involves a struggle usually lasting from about 20 to 60 seconds before submerging. Contrary to popular belief, a struggling victim does not violently panic the way it is often represented in the movies and on TV and seldom does an active drowning victim yell for help. An evaluation of more than a dozen toddler and young child drownings found that yelling for help never occurred. In a study conducted by Fletemeyer, the majority of respondents identified "yelling for help" and "splashing and panicking" as the most likely behaviors associated with drowning (Figure 2.5).

In this same study, 418 movie and TV drowning scenes were evaluated. All but two (99.5 percent) inaccurately portrayed drowning. These scenes typically show

TABLE 2.1
Factors Contributing to over 500 Active Drownings Occurring in the All American Canal over a Ten-Year Period

Factors Contributing to Drowning	Level of Significance
Knowing How to Swim	5
Water Temperature	3
Time of Day	2
Clothing	2
Swimming Aid (i.e., raft, inner tube, plastic jug)	4
Current Velocity	4
Single vs. Group	1
Water Clarity	1

Legal Concepts, Drowning, and Lifeguard Effectiveness

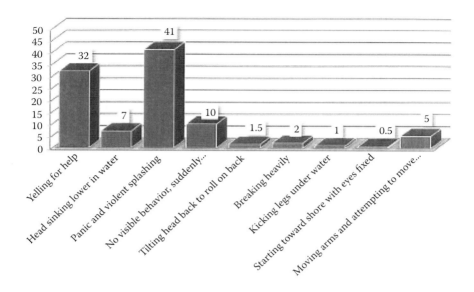

FIGURE 2.5 Perception commonly held by the public about drowning behavior.

a victim violently splashing at the surface, screaming for help, and then quickly disappearing underwater. An analysis of real-life fatal drownings captured on video showed that this never happens, and as mentioned in the above, this is especially the case among young children. In addition, many TV and movie drowning scenes show the victim being revived from cardiac arrest and a cessation of breathing following a few quick rescue breaths performed by the rescuer. In reality this never happens. In a real drowning event, first the victim begins to hyperventilate, and consequently, the head becomes lower and lower in the water while the mouth is at the water level. In a real-life drowning event, the victim is usually facing toward the shore or in the direction of the pool deck. The victim is usually attempting to move toward safety while trying to keep his or her head above the water but with little or no progress. Sometimes this behavior is called "attempting to climb the ladder." The body is usually nearly vertical in the water column attempting to move forward. If the victim is a non-swimmer or poor swimmer, making forward progress is nearly impossible. Maintaining a vertical position often accelerates the drowning process because according to the Archimedes principle, an object in the water is buoyed by the equal amount of water that it is displacing. Consequently, a human body, no matter what its mass, will start sinking faster if maintained in a perpendicular position to the surface as opposed to being maintained in a parallel or near-parallel position (Figure 2.6).

In some cases, a victim will attempt to roll over on his or her back and float but is usually not successful. Eventually the victim begins to tire and begins to submerge. Sometimes the victim will submerge, surface, and re-submerge several times before submerging for the final time. As the victim submerges, he or she attempts to hold his or her breath. Once the victim is underwater and when he or she cannot hold their breath any longer, the victim aspirates and swallows water (Stage 2). As water enters

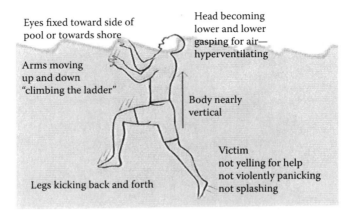

FIGURE 2.6 Figures demonstrating some of the typical behaviors and characteristics of a drowning victim before he or she submerges underwater.

the lungs, the tidal volume of air in the lungs is lost, making the body more and more negatively buoyant.

Usually after swallowing copious amounts of the water (but not always!), the drowning process accelerates as the victim's blood chemistry begins to drastically change. This is called *acidosis*. Although times vary, it usually takes from one to two minutes before the victim becomes unconscious. Breathing stops while the heart usually continues to beat. Before stopping, the heart begins to fibrillate. Typically this lasts between two to six minutes and permits the opportunity for the victim to be resuscitated providing that CPR and defibrillation with an AED begin quickly. CPR usually only sustains life by keeping the tissues oxygenated. In the majority of cases, resuscitation is successful in four minutes or less and in some cases can be successful several minutes thereafter. Once breathing and the heart stops, the patient suffers *clinical death*. Clinical death progresses to biological death at the onset of the deterioration of cellular functions and when there is no longer brain activity. Note that there is debate among experts in regard to "dry drowning." If it does in fact occur, it is exceedingly rare.

Passive/Silent Drowning

A passive drowning involves a secondary event responsible for suddenly rendering a victim unconscious and submerging underwater without any signs of distress or struggle. Some of the secondary events responsible for a passive drowning include the following:

1. Shallow water blackout
2. A sudden seizure
3. Unconsciousness from drugs or alcohol
4. Electrocution
5. A severe trauma accident (i.e., a spinal injury)
6. Cardiac arrest

Cold Water Drowning

When drowning in cold water at least 70°F or lower, the mammalian diving reflex may be triggered. This reflex is responsible for shunting the blood from the extremities and directing it to vital organs, including the heart and brain. The mammalian reflex is responsible for *Bradycardia*, a slowing of the heart rate by up to 50 percent in humans.

In very cold or freezing water reflex reactions can be lethal, killing up to 70 percent of people within 15 to 30 minutes. In cold water a drowning gives rise to cold shock and a combination of uncontrolled gasping and an increase in blood pressure, resulting in possible cardiac arrest. This is followed by the rapid loss of control of the bodily functions needed for swimming.

When understanding drowning in cold water, it is important to consider the few rare but documented cases usually involving a young child that has fallen through the ice or into very cold water. In these rare cases, a young victim has been revived following a period or more than 20 minutes of being underwater. The longest time for a cold water submersion is 66 minutes (Bolte et al., 1988).

Shallow Water Blackout

Shallow water blackout may be a leading cause of death. However, the number of deaths that can be attributed to shallow water blackout is not precisely known because these deaths are often misdiagnosed and reported as a general drowning. When a medical examiner rules a shallow water death as "drowning," it masks the real cause, which is "hyperventilation combined with competitive, repetitive breath-holding." For this reason, shallow water blackout is not well-known or understood by many of those who are most at risk. When oxygen levels fall to critical levels, blackout is instantaneous and frequently occurs without warning. Most of the time, underwater swimmers have no clue that they are about to be rendered unconscious and that they will be vulnerable to death within minutes.

The physiology of shallow water blackout is fairly complicated, which makes explaining it to a jury difficult. Shallow water blackout occurs because of the lower-than-normal carbon dioxide and low oxygen levels. Unconsciousness occurs when O_2 levels in a swimmer are too low. What triggers breathing is high carbon dioxide, not low oxygen as one might think. Hyperventilation done prior to breath-holding lowers the CO_2 abnormally so that one can hold his or her breath longer. However, one may experience shallow water blackout even without hyperventilation before breath-holding. The primary cause of shallow water blackout is a lack of oxygen reaching the brain. The CO_2 levels may be high, as in extreme exertion/exhaustion, or low, as in hyperventilation. In each case shallow water blackout happens. However, with low CO_2 levels, our bodies are robbed of their built-in mechanism to protect us and tell us to breathe before unconsciousness happens. One then basically "blacks out," usually near the water's surface. For some, a victim's lungs will take on water, leading to drowning, while in others, victims simply suffocate or die due to other causes brought on by the breath-holding.

Victims who survive shallow water blackout report that it happens without any warning of its onset. In fact, because of the hypoxia and detached mental state, one can feel euphoric and empowered to continue breath holding. Unlike regular drowning where there can be six to eight minutes before brain damage and death (biological death), there are only about two-and-a-half minutes before brain damage and death with shallow water blackout because the brain has already been oxygen-deprived, coupled with warm water (as in swimming pools), which hastens brain damage.

Shallow water blackout can occur in any pool, bath, spa, lake, ocean, or body of water when breath-holding, regardless of water depth. Even if lifeguards are on duty, there is still a great risk because it is difficult to recognize from above the water. Nevertheless, as trained professionals, a victim experiencing shallow water blackout should never go undetected by a lifeguard practicing the 10/20 rule. To avoid shallow water blackout, the following recommendations should be followed:

- Do not practice prolonged breath-holding.
- Never swim alone.
- A coach or fitness instructor should never encourage breath-holding competitions of exercises.
- Never hyperventilate more than two or three times.
- In guarded pools and beaches, lifeguards should be vigilant for breath-holding games and should take active measures to prevent them.

Toddler Drowning

The courts define a toddler as a child between the ages of one and four years old. In this age group drowning represents the second leading cause of accidental death. By far, the most significant factor responsible for the drowning of a toddler is the lack of responsible adult supervision. In addition, children with autism or a predisposing seizure disorder are especially vulnerable.

When a young child or toddler drowns, the behavior that transpires during the initial phase when the victim is still at the surface is markedly different from the behavior of an older child, teenager, or adult. Older drowning victims often thrash and exhibit erratic behavior before submerging while children and toddlers often remain at the surface only momentarily before silently submerging and without exhibiting any of the drowning behaviors listed earlier in this chapter. This behavior is often missed by an adult caregiver. Consequently, "touch" supervision should always be practiced by a responsible adult while a child is in and around the water.

Drowning in hot tubs and bath tubs are fairly common events. Contributing factors to hot tub and spa drownings include the following:

- Hot water exceeding 104°F, which causes drowsiness
- Drug and alcohol consumption
- Using hot tubs and spas alone
- Failure to supervise toddlers and young children
- Entrapment injuries resulting from lack of an emergency shut-off value and anti-entrapment drains

ENTRAPMENT DROWNINGS

Entrapment drownings and injuries are usually the result of clothing, long hair, an arm, or a leg being trapped in a swimming pool's drain as the result of extreme suction created by a circulation pump and being equipped without anti-entrapment drains.

The Virginia Graeme Baker Pool and Spa Safety Act was passed to prevent entrapment death and injuries. The purpose of this act is to

- Require the use of proper devices, such an anti-entrapment drain covers, through the establishment of a grant incentive program in order to encourage states to enact comprehensive pool and spa laws.
- Educate the public about drowning prevention.
- Establish a federal swimming pool and spa drain cover standard.
- Ensure that public pools are equipped with proper safety devices.

Following the enactment of the Virginia Graeme Baker Pool and Spa Safety Act, public pool requirements included the following:

- Effective on the year after the state of enactment, each public pool and spa must be equipped with anti-entrapment drain covers.
- Each public pool and spa with a single main drain (other than an unblockable drain) must also be equipped with a device or system designed to prevent entrapment, such as a safety vacuum release system.
- Public pools and spas that are open to the public generally (whether for a fee or free of charge) are included. This includes those open exclusively to members of an organization and their guests, those open to residents of an organization and their guests, those open to residents of an apartment building, or those operated by the federal governments. This also requires that every pool or spa drain cover manufactured, distributed, or entered into the stream of commerce in the United States must conform to the entrapment protection standards of the ASME/ANSI A112.19.8 performance standard.

CONSTRUCTION AND WORK-RELATED DROWNING

Often construction projects are located on or near the water. Water- related accidents, including drowning, represent accidents waiting to happen. Consequently, workers should be required to wear life vests. A throw buoy or throw bag should be located on site at a location permitting use at a moment's notice. In some cases, where construction is located over the water, barriers should be established or workers should be required to wear safety lines.

SCUBA AND SNORKELING DROWNING

Fatal SCUBA drownings occur more frequently in salt water environments because significantly more divers dive in the ocean. Consequently, when drowning,

FIGURE 2.7 The seriousness of SCUBA diving drowning is elevated by the danger of air embolism when bringing a victim to the surface.

the body experiences the salt water physiological response described previously (see Figure 2.7). Between 1992 and 2003, 947 fatal SCUBA drownings have been reported in the United States. A cardiac arrest was involved in about half of these fatalities and many of these involved "weekend warriors" and individuals who had predisposing medical conditions, such as arterial disease. The Divers Alert Network (DAN) reports that a fatality occurs once every 200,000 dives.

Drownings involving snorkelers are usually novices often with no formal training. Anyone can go to the local drug store, a dollar store, or K-Mart and purchase snorkeling equipment. In the majority of cases, unless purchased at a professional dive ship, the equipment will not fit properly and will contribute to the cause of drowning. In addition, although recommended, snorkelers seldom wear personal flotation vests, also commonly called safety vests. Consequently, lifeguards must be especially vigilant for signs of an inexperienced snorkeler. As mention earlier, snorkeling often involves hyperventilating, leading to shallow water blackout.

POLICE AND FIRST RESPONDERS

Responding to a drowning victim is challenging and often dangerous, especially for any individual lacking lifeguard training and skills and excellent swimming ability. For a nonprofessional, a response can often be life-threatening. For this reason, unless he or she is a good swimmer, an in-water rescue should never be attempted (Fletemeyer, 1999). The record is replete with cases where a well-intentioned Good

Samaritan attempted a water rescue, only to fatally drown. For this reason, multiple drowning events are relatively common, especially on surf beaches. Among professionals, including paramedics, EMTs, and police officers, there is a moral and legal duty to respond to a victim who is drowning. To stand by and do nothing is inexcusable, even when an SOP dictates otherwise.

Most municipalities have aquatic environments within their jurisdictions. Besides, swimming pools, canals, rivers, ponds, and lakes are commonly found in urban areas. Consequently, all law enforcement and emergency first responder personnel should have water rescue training that provides sufficient knowledge and skills to respond to most rescue situations encountered during their watch without seriously threatening the life of the responding officer. In rare cases, such as if a swift-moving river or stream is located within a particular law enforcement area, technical training might be needed that includes cooperation between fire rescue and police.

It is estimated that at least 50 percent of drownings occur less than 15 feet from shore or near the swimming pool's edge. Consequently, this provides the opportunity to perform rescues without ever having to enter the water. Land-based rescues include the reach-and-throw rescues and can be learned quickly. By simply finding a long tree branch, ladder, or a 2 × 4, it is often possible to extend a reaching object to effect a rescue.

Some police departments place throw bags and lifesaving buoys in the trunk of their patrol cars. Throw bags can be used successfully to rescue a victim more than 50 feet from shore, provided that the victim is conscious and able to grab the bag at one end. When throwing a bag to a victim, the bag should be thrown over and beyond the victim's shoulder and then the bag should be slowly drawn into the victim's grasp. Once grabbing the bag, the officer should then slowly and deliberately retrieve the victim to shore while always encouraging him or her to hold tight.

In some cases, water entry rescues might be necessary but should only be attempted if the officer has sufficient swimming skills to make a rescue without seriously threatening personal safety. SOP manuals should carefully define what should be or should not be done when an officer or first responder is faced with making a rescue in a life-or-death situation. Note that several complaints have received national publicity in cases where a law enforcement officer stood by and did nothing while a victim was drowning.

LIFEGUARDS

Most in-water rescues are performed by lifeguards. In early history, a lifeguard's primary duty was to perform rescues. However, beginning in the early 1970s, this duty changed to the prevention of drowning events. Consequently, the relative number of rescues performed by lifeguards has declined. The importance of lifeguards cannot be underestimated. The CDC was responsible for conducting the first and only study focusing on the effectiveness of lifeguards in reducing drownings. This study resulted in a report published in 2001 titled "Lifeguard Effectiveness: A Report of the Working Group."

The USLA reports that there is a 1 in 18 million chance of drowning in the ocean when a lifeguard is present. Despite the effectiveness of lifeguards, drowning still sometimes occurs in their presence. Schwebel et al. (2010) reports that roughly one-third of drowning deaths in the United States, over 500 individuals annually, occur in lifeguarded pools. It has been reported that an important underlying reason why people drown in guarded pools is that lifeguards have not been properly trained to identify the behaviors typically associated with an active drowning or may be distracted from watching the water. In the most egregious cases, lifeguards have their backs to the water, read while on duty, and engage in long social conversations with bathers (see Figure 2.8).

Below represents a list of factors that may be regarded by the courts as the proximate or contributing cause of a drowning event occurring at a guarded swimming pool, water park, or beach.

- Lifeguard failed to use 10/20 proper surveillance method.
- Lifeguard failed to use proper CPR techniques and failed to use them in a timely manner.
- Lifeguard failed to properly extricate a bather with a suspected head or spinal injury.
- Aquatic environment failed to have an appropriate number of lifeguards to provide proper surveillance and zone coverage.
- Lifeguard is assigned other duties that are not related to lifeguarding (i.e., handing out towels, washing the pool deck, maintaining the beach).
- Lifeguard is distracted due to cell phone use.
- Lifeguard is engaged in reading.
- Lifeguard is distracted because of socializing with patrons.
- Lifeguard fails to maintain active certification.
- Lifeguard leaves pool or zone of responsibility unguarded while going to lunch or to the bathroom.
- Lifeguard fails to observe accepted practice of never "babysitting" young children.
- Lifeguard fails to observe drowning behavior.
- Lifeguard allows breath-holding activity, resulting in a victim suffering from shallow water blackout.
- Lifeguard fails to stop a swimmer or swimmers partaking in risk-taking behavior (i.e., diving into shallow water).
- Lifeguard fails to identify a bather that is intoxicated.
- Lifeguard fails to clear the pool or beach during an electrical storm.
- Lifeguard allows the bathing load to be exceeded.
- Lifeguard fails to close a pool with poor viability.
- Lifeguards are not given time to take periodic breaks.

In most states, public pools are required to have certified lifeguards. The American Red Cross certifies the majority of swimming pool and water park lifeguards while the YMCA of the USA, Ellis and Associates, Starfish, and the America Lifeguard Association certify most of the remainder. Through a memorandum of understanding, the United States Lifesaving Association (USLA) provides training and certifications to open water, surf, and beach lifeguards.

Legal Concepts, Drowning, and Lifeguard Effectiveness

FIGURE 2.8 Despite the effectiveness of lifeguards, drownings often occur in their presence. In these photos, the lifeguards are not watching the water. Consequently, their inappropriate actions invite a drowning event.

Lifeguard training must be a continued process, with recurrent training focusing on rehearsing rescue scenarios and practicing emergency first aid. In most cases, recurrent training programs are conducted daily and include conditioning. Many lifesaving organizations include recurrent training in the SOP manual. Training should be logged for future reference and may be helpful defending a lawsuit.

In all but one certification, sixteen is the minimum age to be a pool lifeguard. The one exception is the America Lifeguard Association that sets the minimum age at fifteen. In addition, in all certification programs, a high level of swimming proficiency and being in good health are prerequisites. The following is a summary of requirements of the various lifeguard certification programs.

YMCA

Requires an American Red Cross CPR for Professional Rescuer Certification and a First Aid Certification, generally at the Emergency First Responder level. In addition, the YMCA programs requires the following:

- Must be sixteen years old by the end of the class
- Must be able to tread water for two minutes
- Must be able to swim 100 yards of the front-crawl stroke
- Must be able to swim 50 yards each of the crawl stroke with head up, sidestroke, breaststroke with head up, and inverted breaststroke kick with hands on stomach
- Must be able to perform a feet-first surface dive in 8–10 feet of water and then swim underwater for 15 feet
- Must be able to perform a series of tasks given by the instructor to demonstrate listening and scanning ability as well as stamina and endurance

The course includes:

- Classroom hours: 14 hours and 30 minutes
- Water/pool hours: 13 hours

AMERICAN LIFEGUARD ASSOCIATION

The American Lifeguard Association's minimum age is fifteen. To pass the course requires the following:

- Must be able to swim 300 yards continuously using these strokes in the following order:
  - 100 yards of the front crawl using rhythmic breathing and a stabilizing, propellant kick, Rhythmic breathing can be performed either by breathing to the side or to the front.
  - 100 yards of the breaststroke using a pull, breathe, kick, and glide sequence.
  - 100 yards of either the front crawl and breaststroke.
- The 100 yards may be a combination of front crawl and breaststroke.

- Starting in the water, swim 20 yards using front crawl or breaststroke, surface dive 7–10 feet, retrieve a 10-pound object, return to the surface, swim 20 yards back to starting point with the object, and exit the water without using a ladder or steps—all within 1 minute and 40 seconds.

In addition, the America Lifeguard Association requires students to be proficient in rescue skills, neck and back injury skills, CPR/AED skills, and first aid skills.

AMERICAN RED CROSS

The America Red Cross certification requires that a candidate must be at least fifteen years of age and must pass certain swimming requirements. The certification remains current for two years. The course includes:

- Rescue and surveillance skills
- First Aid, CPR, and AED training
- Lifeguarding and shallow water training

UNITED STATES LIFESAVING ASSOCIATION (USLA)

The United States Lifesaving Association is the agency charged with training and certifying surf (ocean) lifeguards. The USLA set swimming and skill standards at a higher level than the standards set by the American Red Cross and YMCA for swimming pool lifeguards. The USLA certification prerequisites include the following:

- Age—Minimum of sixteen years
- Swimming ability—Demonstrate an ability to swim 500 meters over a measure course in 10 minutes or less
- Health and fitness—Possess adequate vision, hearing ability, and stamina to perform duties of an open-water lifeguard as documented by a medical or osteopathic doctor
- First Aid Certification—Certified as having successfully completed a first aid course accepted by the federal government or by the state government
- CPR certification—Currently certified as having successfully completed a course providing one person adult, two person adult, child, and infant cardiopulmonary resuscitation including obstructed airway training

A few other organizations provide industry-accepted training and certifications. In most cases swimming pool certifications require that the participant to be at least fifteen years old and must demonstrate swimming proficiency.

REFERENCES

American Red Cross. (1995). *Lifeguarding today.* St. Louis, Missouri: Mosby Lifeline.
Branche, C., Brewster, C., Espino, M., Fletemeyer, J., Gato, R., Gould, R., Keshlear, B., Mael, F., Martinzez, C., Pia, F., and Richardson, W. (2001). Lifeguard effectiveness: A report of the working group. Centers for Disease Control, Injury Prevention.

Fletemeyer, J. (1995). The art and science of drowning recognition. *Responder*, 2(4):12–17.

Fletemeyer, J. (1999). Water rescue methods for Emergency Service Providers, in *Drowning New Perspectives in Intervention and Prevention*, pp. 87–110. New York: CRC Press.

Fletemeyer, J. (2008). Final report—Drowning occurring in American Canal. U.S. Government Document, pp. 1–66.

Fletemeyer, J. (2014). The difference between clinical death vs. biological death. Paper presented at NSPI Drowning Conference, Portland, Oregon.

Fletemeyer, J. (2017). What really happens when someone drowns? *Aquatics International*. April 4.

Fletemeyer, J., and Dean, R. (1994). The anatomy of a dangerous beach. Sea Symposium No. 2, Panama City Beach, Florida.

National Center for Health Statistics. (2000). Vital statistics of the United States, mortality, underlying multiple causes of death. Public files. Hyattsville, Maryland: HCHS.

Schwebel, D. C., Jones, H., Holder, E., and Marciani, F. (2010). Lifeguards: A forgotten aspect of drowning prevention. *Journal of Injury and Violence Research*, 2(1):1–3.

United States Lifesaving Association. (ed. 2000). USLA Open Water Lifeguard Agency Certification Program. Huntington Beach, California.

3 Principles of Aquatic Risk Management from a Legal Perspective

John R. Fletemeyer

CONTENTS

Step 1: The Identification of Risks 55
Step 2: Measuring the Risk 56
    Case Study—Ft. Lauderdale New River Drownings 57
Step 3: Minimizing and/or Eliminating Risks 58
Step 4: Developing an Emergency Response Plan 58
Step 5: Providing In-House and Recurrent Training 59
Step 6: Record Keeping 59
Trespassing 60
    Case Study (Public Pool Drowning) 61
    Case Study (Beach Drowning) 61
    Case Study (Emergency Vehicle Running over a Pedestrian) 62
    Case Study (Movie Night and Swimming Pool Drowning) 62
    Case Study (Multiple Drownings at Lake) 62
References 62

One of the objectives of this book involves identifying aquatic safety hazards and the legal, ethical, and "reasonable" responses to them. In many cases responses to certain hazards are clearly defined by law. For example, the Virginia Graeme Baker Pool and Spa Safety Act establishes a legal requirement for pools to have anti-entrapment drains and for semi-public pools to have surrounding barrier fences with a self-latching gate(s). Responses to many hazards are not required by a legal duty. However, in the absence of a legal duty, a reasonable response is necessary based on the level of risk that a hazard is perceived to represent. In many cases when serious injuries and fatal drownings are involved, what represents "reasonable" is left for juries to decide, but usually with the help of opinions expressed by an aquatic expert.

    In the case of risk management, there are no legal requirements for pools, or for that matter, any of the other aquatic environments discussed in this book to conduct an aquatic risk management assessment. However, it is not possible for any aquatic professional to deny the importance of a risk assessment and that this is the first "reasonable" and necessary step for making the aquatic environment safe whether designed for public use or not.

Risk management is the continuous and ongoing process to identify, analyze, evaluate, and treat loss exposure and monitor risk and financial resources to mitigate the adverse effects of loss. There are several types of losses, including financial, strategic, and operational. The primary objective of aquatic risk management is to prevent injury and loss of life, primarily to a patron or visitor. The other types of loss assume a lower priority.

While the object of risk management practices is to prevent loss, its effectiveness is sometimes limited by human behavioral factors that cannot be controlled even by the most comprehensive and carefully planned program. Risk-taking behavior and horseplay is often responsible for serious accidents and drownings. The lack of responsible adult supervision is often responsible for the drowning of a toddler or a young child. When these behaviors are responsible for an aquatic accident, management and staff should not be held accountable, and in most cases, juries concur. In some cases, these behaviors might be contributing factors and not the proximate cause of an accident or fatal drowning.

Aquatic lawsuits alleging negligence often start with a review of risk management policies. These are often found in a Standard Operating Procedures Manual (SOP). In the worst and most egregious cases, risk management has been ignored by management. In legal terms, "management knew or should have known" the importance of aquatic risk management assessment and practices as it related to the importance of preventing aquatic accidents and drownings. This simplifies developing a theory of negligence because a comprehensive risk management assessment is usually required of every aquatic environment that promotes or encourages recreation activities, especially public pools and beaches. In other aquatic environments, such as semi-public pools, a risk management evaluation is not a requirement but is recommended.

Most aquatic professionals have formal training in risk management principles and practices. Those who do not must rely on a professional risk manager. In some cases, a risk management plan can be developed by a single individual, but in most cases, it requires a coordinated effort by a team of experts. Often an insurance underwriter is a member of this team. Figure 3.1 identifies a risk management team of experts needed to conduct an assessment of a surf beach.

Every aquatic environment has a variety of life-threatening hazards, and consequently there is an implicit duty to respond to them. This response is a comprehensive risk management assessment. The objective of an assessment is to identify within a specific aquatic environment (i.e., a beach or swimming pool) all the aquatic hazards that directly or potentially impact human safety, evaluate (and prioritize) their relative danger, and finally, identify the appropriate response to them. Several times in other chapters, it has been pointed out that many new and unexpected hazards often occur, especially in open water environments. Consequently, this is the reason why risk management must always be viewed as an ongoing process. Once hazards are identified, there are two typical responses:

- The hazard must be eliminated.
- If the hazard can't be eliminated, then the hazard must be minimized by providing effective warnings.

Principles of Aquatic Risk Management from a Legal Perspective

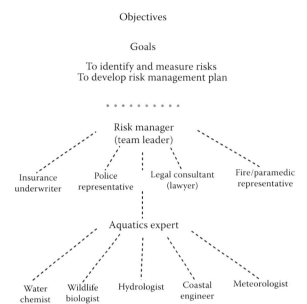

FIGURE 3.1 Identifies a risk management team of experts needed to conduct an assessment of a surf beach.

Warnings serve two vital and but not always mutually exclusive functions:

- Warnings serve as a reminder about a specific hazard.
- Warnings identify a hazard that is previously not known and warns individuals about the inherent dangers associated with the hazard.

There are several types of warnings. Written warnings posted on signs are the most common but not necessarily the most effective. Warning logos are often used, such as the NO DIVING logo commonly seen stenciled on the sides of swimming pools (see Figure 3.2). Research by Fletemeyer identified several factors responsible for reducing the effectiveness of sign warnings. Behavior compliance percentage is the measure of effectiveness and may be different among different age groups and perhaps even among different cultures. For example, teenage boys represent a segment of the population that often intentionally ignore warnings.

Why warnings fail (see Figure 3.3):

- They do not comply with ANSI standards and are not be effective because they are not designed properly. ANSI identifies standards for the color, shape, and size of the sign. These standards are similar to the standards used for highway signs.

FIGURE 3.2 Warnings with accompanying logos are found at all semi-public and public pools. Warning logos are important because guests might not be fluent in English.

- Signs proliferate, resulting in a problem known as "sign pollution."
- Signs are frequently vandalized with graffiti.
- Signs are often confusing and ambiguous.
- Signs may be written in a language not understood.
- Signs become worn and faded, sometimes making individuals feel that they no longer apply to the hazard.

Sometimes colored flags are used on ocean beaches to convey information about beach conditions and warnings. When a blue flag is flown, the intention is to warn the public that man-o-war and stinging jellyfish may be present. Because flags represent a symbolic language, the "receiver" must know what the symbol is intended to represent. In a study conducted focusing on the effectiveness of flag warnings, it was discovered that most bathers were confused about the meaning of a red and a double red flag (see Figures 3.4 through 3.6). This was especially prevalent among tourists. Added to the confusion is that flags are incorrectly displayed and there is often a time delay when "flying" the appropriate flag. On a beach in the Florida panhandle, it sometimes takes more than

Principles of Aquatic Risk Management from a Legal Perspective

FIGURE 3.3 The effectiveness of signs is diminished for several reasons. In these three photos the effectiveness is diminished because of sign pollution, being vandalized, and by being faded by weathering.

FIGURE 3.4 A flag system is commonly used to represent beach conditions, but because flags represent symbolic language, a person must have previous knowledge about the meaning of the different colors of flags. In a study conducted by Fletemeyer (1998), a significant percentage of bathers were not able to determine the meaning of a "red" and "double red" flag.

twelve hours before the correctly colored flag is flown. In addition, many people are colorblind. Consequently, they are not able to distinguish between a danger "red" flag and a "green" safe condition flag.

Aquatic risk management must be regarded as part science and part art and cannot be made effective by training alone. What makes aquatic risk management challenging relates to the fact that hazards often appear quickly and unexpectedly. Examples include sudden electrical storms, the appearance of a rip current or a "smack," or concentration of Portuguese man-o-war floating ashore as strong onshore winds begin blowing.

Implementing a risk management plan following a comprehensive assessment discussed previously, although a straightforward process carefully spelled out in risk management books, should only be attempted by an individual with proper training and experience. Ignoring or forgetting one of the following components could easily be regarded as negligence and become the subject of a negligence complaint, usually and most likely identified as a contributing factor.

Principles of Aquatic Risk Management from a Legal Perspective 55

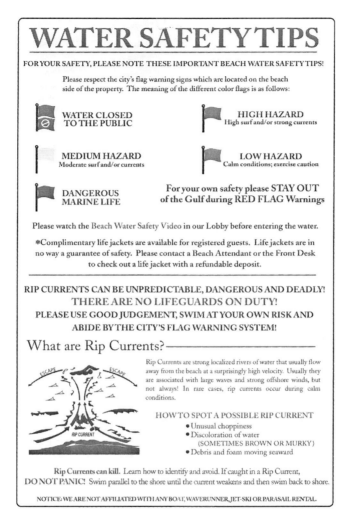

FIGURE 3.5 This brochure is handed to hotel guests and distributed to bathers by lifeguards to explain the meaning of the colored flags.

STEP 1: THE IDENTIFICATION OF RISKS

Every aquatic environment has many hazards that translate to safety risks. In the case of open water environments, hazards (risks) may be seasonal or weather-related. For example, the risk of being bitten by a shark in Florida is significantly greater during the fall when sharks migrate along the coast; the risk of being caught in a rip current is greater during the winter months when surf conditions are usually at their worst.

Swimming pools also have a long list of risks; however, unlike open water environments, risks are easier to identify and respond to. In addition, there are risk management plans already designed for swimming pools that can serve as a guide for other semi-public and public pools.

FIGURE 3.6 Adding to the problem is ambiguity: flags are displayed that are confusing. In this photo both red and yellow flags are being flown.

Open water environments that are privately owned and controlled often represent a special challenge because water represents attractive nuisances where young people often trespass onto the premises. These open water environments often include, ponds, quarries, and canals. They are not intended for recreation but sometimes attract bathers, especially teenagers. At the very least, a risk management plan should include "NO TRESPASSING" signs, and in some cases, a fence barrier. This becomes crucial if one of these environments is located close to an urban area. If owners become aware that trespassing commonly occurs, then more proactive measures might be necessary.

STEP 2: MEASURING THE RISK

Risks have different levels of seriousness in regard to how they potentially impact public safety. The risk of slipping on the pool deck versus the risk of diving into shallow water has a different risk level because a slip-and-fall injury is not usually life-threatening whereas diving into shallow water is. Sometimes during the course of a risk assessment, the level of risk is ranked from 1 to 5, with the greatest risk ranked higher. In cases of a slip-and-fall accident, this risk might be awarded a ranking of 1 or 2 while the risk of a shallow water diving accident might be awarded a 5.

There are various methods to measure risks. One effective method relies on accident statistics. For example, accident statistics reveal that drowning among toddlers and children is a common occurrence when children are left unsupervised by a responsible adult. Consequently this risk deserves a "high" ranking. High-level risks should be given the highest priority in regards to devoting time and money to mitigate. In some cases, the level of risk assessment might change, and this is why risk management must be regarded as a continuous process.

The following is an example of a case that identified a drowning risk and the risk management response that resulted.

Principles of Aquatic Risk Management from a Legal Perspective

CASE STUDY—FT. LAUDERDALE NEW RIVER DROWNINGS

The New River runs through the center of Ft. Lauderdale from the Everglades to the ocean. Tides create strong incoming and outgoing currents. Along several miles of the river, there is a high-profile seawall, making it exceedingly difficult to climb out of the river should someone accidentally fall in. Until recently, drowning in the New River was considered an unlikely event and assigned a low level of risk by city officials. Then, over a period of six months, two adult men fell into the New River and drowned. A month later, a woman fell into the river while attempting to save her dog and nearly drowned. As a result of these three drowning events, the level of risk was elevated to a "high" level, and in response, several lifesaving rings were placed along the river (see Figure 3.7). In addition, several training programs were conducted by the local fire/rescue department to practice extricating a drowning victim from the river.

FIGURE 3.7 Following two fatal drownings and one near-drowning in the New River, city officials elevated the level of risk and responded by placing several lifesaving buoys in strategic locations along the river.

STEP 3: MINIMIZING AND/OR ELIMINATING RISKS

As mentioned earlier, when a risk is identified, there are two responses. The first and most desirable is to eliminate the risk. If this is not possible, then a risk must be minimized by providing warnings. Failure to warn about hazard may be regarded as a breach of duty.

Perhaps the best example of eliminating a risk involves removing a diving board. Several decades ago, most hotel pools had diving boards. Because diving boards have been responsible for numerous serious injuries, most boards have been removed, often at the insistence of insurance underwriters.

An example of minimizing risk involves the installation of anti-entrapment drains. The Virginia Graeme Baker Act mandates that all older pools be retrofitted with anti-entrapment drains and that all newly built pools have these installed during the construction process. Consequently, the risk of an entrapment injury has been significantly reduced.

STEP 4: DEVELOPING AN EMERGENCY RESPONSE PLAN

The objective of an emergency response plan is to apply industry-accepted protocols that respond to an anticipated emergency. In the majority of cases, an ERP appears in an SOP. It carefully defines the action that should be taken by staff during an emergency. The *Aquatic Risk Management* publication by the National Swimming Pool Foundation states (p. 31):

> The facility must have an aquatic emergency response plan that the staff are familiar with. The (ERP) is a detailed plan of the duties of staff during emergencies and should be practiced often to ensure that everyone works together as a team during an emergency.

There are literally hundreds of examples requiring an emergency response. The following includes a few:

- A victim in cardiac arrest
- A victim suffering a spinal injury
- A victim of a shark attack
- A catastrophic accident—boat or plane accident
- A violent disagreement between patrons resulting in physical harm
- A bomb scare
- An electrical storm
- A chemical or blood contamination incident

To reiterate from the above, developing an emergency response plan should be a team effort and should take the following into consideration.

- Types of emergencies most likely to occur.
- The training and experience of the staff.

Principles of Aquatic Risk Management from a Legal Perspective   59

- Layout of the facility. Consideration must be made as to the best and most efficient way for emergency first responders to gain access and then to expeditiously remove a victim from the premises.
- Available equipment to assist in responding to emergencies.
- Chain of command. Identifying and defining the expectations and requirements of staff when involved in an emergency.

STEP 5: PROVIDING IN-HOUSE AND RECURRENT TRAINING

Risk management and an emergency response plan must be practiced. It is incumbent that the aquatic director/supervisor schedule regular recurrent training.

STEP 6: RECORD KEEPING

It is crucially important to maintain accurate records. In the case of swimming pools, records should include chemical values (at the time taken), attendance, weather conditions, water clarity, pool maintenance, and responses to hazards (i.e., closure resulting from an electrical storm).

The failure to develop a risk management plan may result in injury of death of a patron and may represent a contributing claim of negligence. Trying to defend a suit alleging negligence due to the lack of a risk management plan becomes problematic, especially involving a public pool or beach.

Risk management is an ongoing process. In the majority cases, an aquatic risk management program requires continuous and frequent inspections of the premises or facility. When problems are identified, they must be responded to without delay. If this is not possible and the risk is high, and if a pool is involved, it should be closed until a solution to the problem is found.

In semi-public and public swimming pools, daily inspections are the norm. Usually the pool is inspected by an individual with Certified Pool Operators training. Typically a pool is inspected for the following:

- Pool chemistry (sometimes several times a day).
- Water visibility (whether one is able to see clearly the bottom of the deepest part of the pool).
- Drains are functioning properly and not clogged.
- Ladders are stable.
- Pool decks are clear of algae and debris (not slippery).
- Gates open and close properly.
- Bathing load is not exceeded.
- Scuppers are working properly.
- There is no exposed electrical wiring.
- Locker rooms are clean and ready for use.
- Warning and rule signs are displayed properly and not obstructed.
- Barrier fence is without damage or compromise.
- Gates are functional and working properly.

- Emergency equipment (including AED, first aid kit, spinal board, cervical collars, whistles, and lifeguard buoys) is ready and in good order.
- Lifeguard chairs are checked.
- Float line is checked.
- If the pool has starting blocks, they are checked to ensure stability and that the surface is not slippery.
- If the pool has diving board(s), they are checked to ensure safety and that the surface is not slippery.
- If it is a semi-public pool, it is checked to ensure there is a shepherd's hook and lifesaving buoy strategically placed.
- If it is a semi-public pool and a surveillance camera is used, it is checked to ensure functionality.
- If the pool has a dedicated 911 number, the phone is checked to ensure it is operational.
- If the pool is operated at night, pool lighting is checked and bulbs are replaced as necessary.

On guarded beaches, a risk management plan might include the following:

- Identify potential hazards including natural and human-related.
- Remove hazards or provide warnings.
- Check lifeguard towers and maintain them (i.e., clean windows).
- Enter the water (periodically) to determine the presence of in-water hazards, such a rip currents.
- Check communications, including radios and phones.
- Check showers and bathrooms.
- Check patrons for alcohol consumption.
- Check for rule enforcement and children who are improperly supervised.
- Check emergency equipment, including AED, first aid kit, buoys, paddle boards, spinal board, oxygen, and so on.
- Check and maintain emergency vehicles, including Jeeps, trucks, ATVs, and IRBs. Ensure vehicles are filled with gas and fluids.

TRESPASSING

One of the challenges of risk management is trespassing. Stated throughout this book is the fact that aquatic environments are regarded by the courts as "attractive nuisances." Simply stated, the doctrine of "attractive nuisance" states that a landowner or possessor cannot maintain his or her property in such a way or knowingly allow a condition to exist that is dangerous to children who are unable to appreciate the danger and who are known to frequent the premises (Stern et al., 1979). The record is replete with cases where children have wandered onto a property with a pool or pond and then drowned. In the majority of cases, the owner has been held liable if it was proven that he or she did not take reasonable measures to prevent trespassing by children. It should be noted that the doctrine and definition cited above does not apply to teenagers and adults and is often applied differently by courts in different states.

The bottom line is that the owner of a property is expected to exercise reasonable care to prevent trespassing by children and this responsibility is elevated when the water environment is located in close proximity to a school, housing development, park, or playground where children are known to frequent. In regards to pools, reasonable care involves a barrier fence that can include the side of a house that meets certain specifications. In open water environments such as a pond, quarry, or canal, reasonable care becomes more problematic. In remotes areas, no trespassing signs might be regarded as sufficient response. However, in urban areas, a barrier fence might be viewed by the courts as a requirement.

The following are several case studies involving either a serious accident or fatal drowning. It identifies that risk management response taken following the incident. In all of these cases, a risk management assessment conducted prior to the accident may have avoided these accidents.

Case Study (Public Pool Drowning)

At a public pool located in the NE, the water had become cloudy. Consequently, the bottom of the pool could not be observed from the pool deck. Instead of closing the pool until the water cleared, the deep end was roped off and bathing was allowed only in the shallow end. The following day, while a lifeguard was vacuuming the pool, the guard felt the vacuum strike an object in the deep end. The object was the body of a lady who had drowned the previous day. A lawsuit was filed alleging gross negligence and the failure of a policy to close the pool when water visibility was bad.

It was later determined that there was no risk management plan requiring pool closure when water conditions prevent seeing the bottom of the pool. It was also noted that it is an industry practice to close the pool when water visibility becomes so bad that that the pool bottom cannot be seen.

Case Study (Beach Drowning)

At a state-owned beach in Florida, there were rip currents on Memorial Day. The park rangers failed to fly a red flag and continued to fly a yellow flag that had not been changed to accurately reflect the sea conditions. In addition, the park rangers had an opportunity to warn guests about rip currents when they paid the entrance fee. Early in the afternoon, a thirteen-year-old boy was caught in a rip current and drowned. A lawsuit was filed that claimed that there was a lack of a risk management plan that required the flag to be regularly changed and that patrons should have been verbally warned, as they had paid a user fee to a park ranger who was on duty and who was aware of the hazard. Although the park staff was guilty of negligence for not warning patrons about risk currents, the suit was dropped because of sovereign immunity and a recovery "cap."

CASE STUDY (EMERGENCY VEHICLE RUNNING OVER A PEDESTRIAN)

At a crowded public beach, a supervisor parked his emergency vehicle on the sand for several minutes. In the meantime, a woman laid her blanket down in front of the vehicle and lay down. As the supervisor started to drive away, he ran over the woman, causing serious facial injuries and two broken ribs.

The municipality where this accident occurred was successfully sued and damages were recovered by the plaintiff. In response to this injury, vehicles were equipped with emergency beepers and were restricted from driving on the beach during crowded days. In addition, a revised risk management plan required emergency vehicle operators to first check around the vehicle before driving.

CASE STUDY (MOVIE NIGHT AND SWIMMING POOL DROWNING)

At a country club located in Texas, a swimming pool movie night was a featured event. The pool was protected by lifeguards. Children were given floats, then the pool lights were turned off and the movie was started. Following the movie, the pool lights were turned on and the body of a young boy was discovered on the pool bottom.

The country club was sued with the proximate cause citing a violation of lighting standards which prevented lifeguards from monitoring the pool. Outdoor pool movie night was subsequently terminated. This case resulted in national publicity and was responsible for ending movie night at most pools.

CASE STUDY (MULTIPLE DROWNINGS AT LAKE)

At a lake in Oregon with a beach designated for bathing, a drought caused the water to recede, resulting in a sudden drop off in the designated bathing area. Consequently, a family of four suddenly found themselves in deep water and drowned. A lawsuit followed, alleging that there were no warning signs or a risk management plan requiring that the lake be periodically inspected for hazards such as a sudden drop-off that was responsible for claiming the lives of the four victims. At the time of this book's publication, this case is ongoing.

REFERENCES

Fletemeyer, J. (2004). An overview of beach safety in Panama City with short-term and long-term recommendations for improvement. Report funded and published by the Panama Beach and Bay County Tourist Development Council.

Fletemeyer, J. (2012). The effectiveness of aquatic warning signs. Paper prevented at the NSPI National Aquatics Conferences, Portland, Oregon.

Johnson, K. (1999). *The Encyclopedia of Aquatic Codes and Standards*. National Parks and Recreation Association, special publication.

National Swimming Pool Foundation (2009). *NSPF Volume 1: Aquatic Risk Management*. National Swimming Pool Foundation Publ.

Stern, J. F., and Hendry, E. (1977). *Swimming Pools and the Law*. Milwaukee, WI: S. and H. Publishing.

4 Evidence Collection and the Daubert Standard

Lori St John

CONTENTS

Evidence Collection 63
The Daubert Standard 69
Endnotes 71

EVIDENCE COLLECTION

Evidence is the key to the successful defense, and prosecution, of all criminal and civil cases. When it comes to examining a death said to be caused by accidental circumstances, evidence, or the lack thereof, becomes even more important. Accidental deaths, especially drownings, can be complicated to litigate. This discussion is limited to such cases.

Evidence, and motive, is the primary focus of any death that appears suspect in nature. To the prosecution and defense attorney, it is used to build or defend a case. To the layperson, it is what aids him or her in understanding the cause of or reasons for a tragedy.

Accidental deaths are generally, by nature, not the subject of intense investigations unless there is the suggestion of foul play. They are always difficult for family and friends to accept because of the sudden and unexpected loss of a loved one. Most are, however, accepted as accidental and the circumstances that caused the death more palpable to understand. Drownings, however, are often different when it comes to investigation and evidence-gathering. When a victim is found in water and no other obvious causes of death are discovered, the medical examiner (and investigative team) presumes the person has drowned. Circumstances, rather than evidence or tests, generally are the basis for the conclusion of an accidental drowning. For the litigator, it is important to differentiate five variations of circumstances that should be examined in taking on one of these cases: (1) A murdered victim later placed in a body of water to make it appear as a drowning; (2) a person who was forcibly held under water by a perpetrator; (3) drowning by accident; (4) an act of suicide; and (5) a medical or otherwise fatal, or near fatal, circumstance that resulted in a fall into the water.

An autopsy is the final determination of the manner of death that led to a drowning. However, it is important to understand and utilize all facts relevant to the demise of the victim even before they are examined. Sometimes things are not always what they seem. Therein lies the importance of the identification, collection, and preservation of evidence. Some evidence will be used in a forensic analysis by experts such as pathologists, toxicologists, or botanists. For example, a pathologist may attempt to determine if the victim died in salt or fresh water. This may become an issue if the deceased is found in salt water, but the water in the lungs is fresh water. Depending upon the circumstances of the death, this would be useful if there are any suspicions of foul play. An experienced pathologist is essential to the proper interpretation of findings, supported by adequate histological, biochemical, and toxicological analysis of material taken at autopsy. Equally as important is an investigation into the circumstances that allegedly led to the victim's death. In some cases it is important to simulate a crime scene reconstruction. If the person is said to have drowned when swimming back to a boat with other members of his party, did the investigative team examine the boat for any sign of blood or other evidence leading to foul play? Has a proper and time-sensitive investigation taken place as it pertains to interviewing all witnesses? Have proper and documented interviewing techniques been used to obtain the alleged circumstances related to the demise of the deceased? Has the information obtained through the interviewing process been evaluated and noted for any inconsistencies, probabilities, and reliability? An investigation of the circumstances leading to one's death entails understanding the deceased person's life and abilities (such as if he or she was a good swimmer), relationships with parties close to or at the drowning scene, and any possible motive that may exist to lead to foul play. Due to the nature of a drowning, and the lack of evidence normally attributable to such, your client may be wrongly accused of murder if circumstances are suspicious, or even inconclusive, surrounding the incident which led to the death.

In 2013, a team of seventy-five seasoned and well-respected attorneys, law enforcement officers, scientists, and other criminal justice experts documented their findings in key areas that contribute to wrongful convictions in a report titled "The National Summit on Wrongful Convictions: Building a Systematic Approach to Prevent Wrongful Convictions."[1] One of the four key areas studied was the leveraging of technology and forensic science. Evidence collection and investigation techniques were also evaluated in the study. In the report, experts reviewed the kind of mistakes, omissions, and judgment errors that lead to wrongful convictions. Their findings resulted in the recommendation for better communication, training, protocols, supervision, assessment, and review.[2]

Also in 2013, the National Institute of Justice and National Institute of Standards and Technology with the Office of Law Enforcement Standards released *The Biological Evidence Preservation Handbook: Best Practices for Evidence Handlers.*[3] A technical workgroup focused on biological evidence preservation, discussing the retention, safety and handling, packing and storing, and disposition of biological evidence. The team also reviewed tracking the chain of custody of such evidence and came up with a summary of recommendations outlined in pages 43–45 of the report.[4] The technical workgroup included individuals from the Forensic Services Bureau in the Miami-Dade Police Department, the Evidence Control Unit of the Federal Bureau of Investigation, the International Association for Property and Evidence, the Law Enforcement Standards Office of the Office of Special Programs, and several other agencies. These standards are useful in the collection, preservation, chain of custody, and disposition of evidence. See Figure 4.1 for a photograph

Evidence Collection and the Daubert Standard 65

FIGURE 4.1 A well-organized property and evidence room ensures a proper chain of custody.

of one section of a well-organized property and evidence room in a municipal police department. The biological evidence is clearly marked and either located on a shelf or in the refrigerator, depending on the nature of the evidence.

To ensure the chain of custody, each piece of evidence must be carefully identified, labeled, and processed according to carefully laid-out police policies and procedures. These departmental policies and procedures include the handling process; evidence location (the section of the property and evidence room(s) and the distinct location on the shelf or bin); who authorized and transported the evidence to the laboratory for testing (or the courtroom for trial); and finally, when applicable and authorized by state statute, its disposition. In this case, the evidence is a revolver with bullets. The packaging has been properly sealed and signed over the seal to prevent tampering, and paperwork accompanies the evidence for a proper chain of custody. For police departments to be accredited by the Commission on Accreditation for Law Enforcement Agencies (CALEA), an annual audit of the property and evidence room must be performed. While it can be done in-house, to diminish the appearance of a conflict or impropriety it is wise to engage an outside audit firm or a CPA with specialized knowledge that includes both auditing and evidence matters.

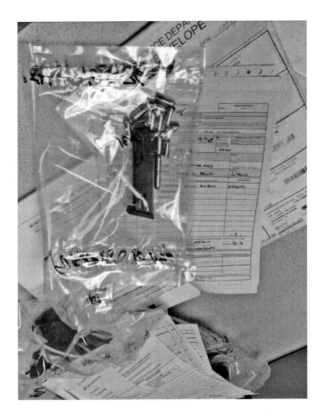

FIGURE 4.2 Evidence must be properly collected, labeled, and stored to retain a chain of custody.

Because other chapters in this book address forensic evidence, I will limit my discussion to just a few types of evidence that one should seek to obtain and examine when investigating or litigating these types of cases (see Figure 4.2). Due to the similarity of some symptoms of a drowning it is difficult to distinguish an accidental drowning from one caused by a medical or drug-related condition.

> Symptoms such as hemorrhages, pulmonary edema, "washerwoman" appearance of the hands and feet, and goose flesh may indicate a drowning. However, heart attacks, drug overdoses, and other causes of death can also cause the same symptoms. Wounds that do not bleed also do not help distinguish the manner of death because they could have been inflicted post or antemortem. A good indicator that the person was alive while submerged is the presence of vegetation and stones from the bottom of the water in the hands of the victim. So difficult is it to determine the cause of death, an innocent person could be prosecuted for murder for an accidental drowning or a suicide.[5]

One thing for certain is the need to perform an on-scene body assessment. This entails examining the location and position of the body and collecting any botanical and all other forensic evidence possible. Once you remove the body from its original location, any conclusions drawn from a forensic evaluation may become suspect and challenged by opposing counsel, suggesting contamination or foul play. So be sure the location and position of the body is documented and photographed, as well as noting and collecting as much potential forensic evidence at the time of the initial discovery of the body.

An on-scene assessment should consist of the following observations[6]:

- Postmortem wandering
- Ocular (eye) changes
- Foam column/foam cone
- Rigor mortis
- Livor mortis
- Algor mortis
- Physical wounding
- Anthropophagy (animal/fish feeding)
- Maggots
- Decomposition
- Skin slippage

An on-scene assessment should also include an evaluation of the area in which the victim was discovered. For example, if in a lake, was the bottom sandy or rocky? In addition, all bruises, abrasions, physical abnormalities, or noticeable wounds should be noted, as well as their exact location. The physical appearance of the victim can assist the coroner in determining the timeframe in which wounds were sustained. Notation of the wounds should be done at the time of recovery to avoid any challenges due to conditions during the transportation to the medical examiner's office. It should be determined whether these wounds occurred before death (antemortem), at the time of death (agonal), or after death (postmortem).

> Positioning of some injuries may indicate the time when they were sustained. Propeller injuries before death would be found on the hands, abdomen, and legs since the victim would be upright in the water, whereas propeller injuries across the shoulders, back of the head and back would indicate postmortem wounding, since the body would be floating in the facedown position. Sharp incised wounds on the palmar region of the hands and outer portion of the forearms may indicate defensive wounds from the victim warding off an attack, whereas injuries to the back of the hands may indicate travel abrasions or other post-mortem injury from the body coming in contact with objects in the water.[7]

In the identification and collection of evidence, a fish/aquatic feeding and the discovery of maggots on the body may be informative in evaluating a drowning case. If the body is exposed to air then flies may lay their eggs on the body. A forensic entomologist should examine maggots collected at the scene. This can also assist in determining the time of death of the victim. A study of the aquatic organisms located in the particular body of water in which the victim was found is important to draw conclusions regarding injuries noted on the body. Small circular marks can be attributed to leech bite marks. Sea lamprey can also cause circular-shaped marks, but generally with a depression of the flesh.[8] Ocean forensics helps scientists determine if a mark on the body is from a predator or if it was from foul play. Circular marks are often related to fish bites.

In addition to the observation of aquatic feeding on the body, the appearance of the body is equally as telling. Based upon the condition of the body it is possible to determine the approximate length of time it was underwater before recovery. During various stages in the water a body will reveal a different appearance. For instance, at a twenty-four-hour timeline there would be pooling of the blood and consequently there would be a significant change in color to the extremities.[9] During the stage of early decomposition, the body will exhibit an ashen color. See Chapter 13 for discussion on decompression of the body after death.

Unfortunately, drownings are often determined based upon negative evidence rather than positive evidence. Largely, unless there are obvious other signs of toxicologic or anatomic findings, the victim is presumed to have drowned. Unfortunately, there are no universally accepted diagnostic laboratory tests for drowning.[10] As such, an investigation should be conducted without the preconceived notion that simply because a body has been found in water the cause of death is drowning.[11] Therefore, there must be detailed examination of the clothing and the deceased's body noting marks such as abrasions, wounds,[12] and so on. During the internal examination of the body, a detailed microscopy should be performed to determine the nature, extent, and stage of any antecedent disease.[13]

> A diagnosis of drowning cannot be made without a complete autopsy and full toxicologic screening, histologic analyses of all organs including the lungs, and the diatom test. The diagnosis of drowning cannot be based solely on the circumstances of the death, nonspecific anatomic findings, and the results of biologic analyses.[14]
>
> If the anatomic and toxicologic observations are equivocal, the opinion that death resulted from drowning is merely presumptive, and death is attributed to drowning only because [sic] autopsy failed to disclose some other cause. Obviously, it is preferable that a diagnosis of death from drowning (indeed, from any cause) be based on positive rather than on negative evidence, but this desirable state of affairs does not always obtain [sic] (Adelson at p. 558).[15]

Moreover, it is essential to obtain a full medical history of the deceased prior to an examination of the body. The medical examiner or police investigators should collect the information and make it known to the forensic pathologist before the autopsy begins. Evidence also includes the past history of the decedent, as in the case of a bathtub drowning.

> A fatal accident in the bathtub is frequently associated with an epileptic attack, an episode of acute coronary insufficiency, and alcohol or drug intoxication. In such a death also minor injuries from a fall may be present. The body should be examined for injection marks and electrical burns. Past history of the decedent must be reviewed (Fatteh at p. 155).[16]

Finally, evidence is the single most important mechanism to resolve the question of whether the death of an individual in a drowning case is accidental or the result of foul play. It cannot be underestimated in the purview of the criminal justice system. Without the proper knowledge, follow-through, and an understanding of all that goes into an investigation of a drowning, the litigator cannot prevail. Equally as important is what was not done in an investigation. Thereto are the answers to what may have transpired. To evaluate and confirm the occurrence of an accidental drowning is one thing, but to get it wrong and accuse one of murder runs the potential risk of a lifetime of injustice.

THE DAUBERT STANDARD

As the litigator prepares for trial, no matter how good the evidence is or is not, you better be armed with a solid and prepared team of experts. Depending on the state in which you are trying your case, you will be challenged on the admittance of expert testimony according to the respective states' standard. Formerly, the *Frye* standard was the standard used to admit testimony by expert witnesses. The *Frye* standard, or *Frye* test, was the general acceptance test used to determine the admissibility of scientific evidence. In 1923, in *Frye v. United States*, the District of Columbia Court rejected the scientific validity of the lie detector (polygraph) because its technology did not have significant general acceptance at the time. The court established the general standard that a scientific principle be sufficiently established to have gained general acceptance in the particular field in which it belongs. Plainly speaking, the court had to decide whether a procedure, technique, or principle in question was generally accepted by a meaningful proportion of the relevant scientific community. Scientific evidence was admissible if it was based on a scientific technique generally accepted as reliable within the scientific community.[17] This standard prevailed in both the federal and state courts for many years, until the Supreme Court held that *Frye* was superseded by Federal Rule of Evidence 702.[18] Over time, *Frye* was challenged, and there evolved the Daubert standard for admitting expert testimony into court.

In *Daubert v. Merrell Dow Pharmaceuticals, Inc.* 43 F.3d 1311 (9th Cir. 1995), the court explained that the federal standard included not only general acceptance, but also consideration to science and its application. Under Daubert, there is a two-part analysis. First, the court "must determine whether the experts' testimony reflects "scientific knowledge," whether their findings are "derived by the scientific method," and whether their work product amounts to "good science." Second, the court must ensure that the proposed expert testimony is "relevant to the task at hand," that is, that it logically advances a material aspect of the proposing party's case. "The Supreme Court referred to this second prong of the analysis as the 'fit' requirement ... the question of admissibility only arises if it is first established that the individuals whose testimony is being proffered are experts in a particular scientific field."[19] Noting the difficulty in applying the first prong of the Daubert standard, the court listed several factors federal judges could utilize in determining whether to admit expert scientific testimony under Federal Rule of Evidence 702:

- Whether the theory or technique employed by the expert is generally accepted in the scientific community.
- Whether it has been subjected to peer review and publication.
- Whether it can and has been tested.
- Whether the known or potential rate of error is acceptable.

The above criteria are said to be relevant in every case; by no means is it also meant to be exhaustive in nature. Instead the court in Daubert read the Supreme Court as guiding them to determine whether "... the analysis undergirding the experts' testimony falls within the range of accepted standards governing how scientists conduct their research and reach their conclusions."[20] In making this determination the court relies upon whether the expert has conducted his or her own research rather than having done so solely for the purpose of litigation. This is said to be the most persuasive basis "for concluding that opinions expressed are derived by the scientific method."[21] If this cannot be done, the court in Daubert notes that the research and analysis supporting the conclusions drawn by the expert can be supported by having been subjected to peer review and publication.[22] An important distinction is drawn between reaching conclusions that are "correct" and those that are based upon the soundness of the methodology. The test under Daubert is the latter.

Under Daubert, the court next considered the second prong of the Rule 702 admissibility inquiry, "Would the plaintiffs' proffered scientific evidence 'assist the trier of fact to ... determine a fact in issue'" (Fed. R. Evid. 702). Daubert stressed the importance of the "fit between the testimony and the issue in the case ... which requires a valid scientific connection to the pertinent inquiry as a precondition to admissibility."[23]

The two-prong test under Daubert is used to determine the admissibility of your expert's opinion. This can strengthen or weaken your case depending on the expert you engage and his or her knowledge and preparation as it pertains to admissibility. It is vital that you walk through the standard with your expert to ensure that upon deposition, and at trial, he or she is aware of what may be asked by opposing counsel in an attempt to disqualify him or her.

Federal Rule 702, which was adopted in 1975 for litigation in federal courts, is currently used by several states. In 1993, the United States Supreme Court held, in *Daubert v. Merrell Dow Pharmaceuticals, Inc.*, that the Federal Rules of Evidence, and in particular Rule 702, superseded the *Frye* standard.[24] Scientific expert testimony is now subject to the Daubert standard in several states (see Figure 4.3). However, there are some states that still utilize the *Frye* standard, or a modification thereof.[25]

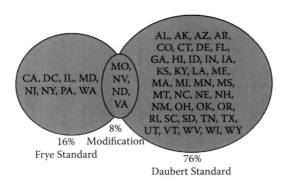

FIGURE 4.3 A summary of state admissibility standards reveals that the majority of states use the Daubert standard.

According to the Rules of Evidence in 2016, state admissibility standards reveals that 76 percent of the states use the Daubert standard, 16 percent use *Frye*, and the remainder use a modification.[26]

Finally, if you are a seasoned litigator or crime scene investigator this chapter may be second nature to your practice. However, for those who are either new to this field or to the practice of law, it is important to know and understand the logistics of evidence collection, preservation, and the chain of custody. Interview techniques are equally as important to master when you, or your investigator, seek to support your case with witness testimony. Moreover, a clear understanding of the standard that governs the admittance of expert testimony in your jurisdiction is vital to your success.

ENDNOTES

1. The National Summit on Wrongful Conviction: Building a Systemic Approach to Prevent Wrongful Convictions. Report from the International Association of Chiefs of Police/U.S. Department of Justice, Office of Justice Programs Wrongful Convictions Summit. August 2013.
2. *Id.*
3. *The Biological Evidence Preservation Handbook: Best Practices for Evidence Handlers.* National Institute of Standards and Technology, U.S. Department of Commerce, National Institute of Justice. April 2013.
4. The Biological Evidence Preservation Handbook: Best Practices for Evidence Handlers, pp. 43–45.
5. Lerskine, K. 2011. An introduction to water-related death investigation, part I. *Evidence Technology Magazine.*
6. *Id.*
7. Armstrong, E. J., and Lerskine, K. 2016. *Water-Related Death Investigation.* Boca Raton, FL: CRC Press.
8. Department of Environmental Conservation, New York State, http://www.dec.ny.gov/animals/6998.html
9. Dr. Fletemeyer. 2018. *Drowning Forensics, Drowning and Accident Reenactments, and Case Research*, p. 258.
10. Moles, R. N., and Barrister, B. S. Networked Knowledge-Medical Issues. Diagnostics of Drowning Cases, Lectures in Forensic Medicine, University of Dundee, Derrick Pounder, http://netk.net.au/Reports/DiagnosticsofDrowning.asp.
11. *Id.*
12. *Id.*
13. *Id.*
14. Payne-James, J., Busuttil, A., and Smock, W., eds. 2002. *Forensic Medicine, Clinical, and Pathological Aspects*, p. 250.
15. *Id.*, p. 558.
16. *Id.*, p. 155; Moles, R. N., and Barrister, B. S. Networked Knowledge-Medical Issues. Diagnostics of Drowning Cases, Lectures in Forensic Medicine, University of Dundee, Derrick Pounder, http://netk.net.au/Reports/DiagnosticsofDrowning.asp.
17. *Daubert v. Merrell Dow Pharmaceuticals, Inc.*, 43 F.3d 1311 (9th Cir. 1995).
18. *Id.*
19. *Id.*
20. *Id.*
21. *Id.*

22. *Id.*
23. *Id.*
24. *Id.*
25. Morgenstern, M. 2017. Daubert v. *Frye*—A State-by-State Comparison, https://www.theexpertinstitute.com/daubert-v-frye-a-state-by-state-comparison.
26. *Id.*

5 Swimming Pool and Spa Safety

John R. Fletemeyer

CONTENTS

Swimming Pool and Spa Safety 73
    Introduction 73
    Hot Tubs and Spas 79
    Toddlers 80
    Risk-Taking Behavior 81
    Secondary Contributing Factors 81
        Swimming Pool Drains and Entrapment Injuries 81
        Swimming Pool Lighting 82
        Swimming Pool Water Clarity 83
        Swimming Pool Decks 84
        Diving Boards 85
        Barrier Fences 85
        Gates 87
        Waterslides 88
        Swimming Pool Design and Engineering 88
        Signage 88
        Depth Markers 90
        Hot Tubs and Spa Signs and Safety 90
        Swimming Pool Bathing Load 92
        Water Chemistry 93
        Safety Equipment 94
        Lifeguard Protection 94
    Overseas Swimming Pool Safety 96
    Other (Special Events, Parties, Shallow Water Blackout, Sexual Predators,
    and Emergency Closures) 96
Reference 97

SWIMMING POOL AND SPA SAFETY

INTRODUCTION

As more and more swimming pools are built and as the middle class expands, Americans have more leisure time to spend bathing in pools and engaging other aquatic recreational activities. Consequently, accidents occurring in and around swimming pools will increase and so will litigation alleging negligence.

Thanks to the Virginia Graeme Baker Pool and Spa Safety Act and nationally inspired directives and model recommendations from ANSI and the Consumer Products Safety Commission (CPSC), state and local municipalities are being urged to adopt more stringent, uniform swimming pool codes and regulations. As a result, most states have adopted codes and regulations to make swimming pools safer. Violations of these codes and regulations often result in penalties ranging from a verbal warning to, in especially egregious cases, a fine and forced pool closure until appropriate remedial action is taken. Although states are being more proactive about swimming pool safety, accidents continue to happen.

The objective of this chapter is to identify some of the factors that are responsible for serious swimming pool accidents and fatal drownings. There is far too much information about this subject to cover in only a chapter. Consequently, information presented in this chapter must be regarded as an overview that nevertheless represents a useful and insightful research reference about swimming pool safety. This chapter is intended for an audience that includes plaintiff attorneys, defense attorneys, risk managers, and insurance underwriters. However, it may also serve as a useful reference for aquatic professionals.

Ultimately, the owner—and in some cases, the owner's delegates, including managers and staff—is responsible for the safety of a swimming pool and the invited guests using the pool. In unusual situations, the courts have ruled that this responsibility may extend to uninvited guests. In commercial pools, including semi-public and public swimming pools, swimming pool safety, maintenance, and operation may be delegated to a third party, such as a swimming pool maintenance company or a private company specializing in providing lifeguarding services, such as Ellis and Associates. A growing trend and one that is primarily economically motivated is that some large community and resort pools are being maintained by outsourcing. Companies such as Ellis and Associates have assumed this role of outsourcing swimming pool maintenance and safety by providing lifeguards.

The CDC as well as many other organizations and individuals believe that most aquatic accidents are preventable. Literature and court rulings support this belief. However, the pool owners are not always to blame. Frequently, the behavior of the swimming pool user is a contributed factor, and in some cases, the proximate cause, of a serious swimming pool accident or fatality. For example, horseplay and alcohol are blamed in about 50 percent of all serious aquatic accidents and in many cases, the lack of responsible adult supervision is the proximate cause of a fatal drowning involving a toddler. Clearly, in these cases where alcohol, reckless behavior, and the lack of adult supervision are involved, the owner is not usually to blame.

Over the years, as bathing and swimming has become more popular, the number of swimming pools has increased dramatically. The exact number of swimming pools and spas in the United States is not precisely known, but it is estimated that there are at least 13 million. The number of pools and spas are listed in the following four categories.

Regarding in-ground pools, the aquatics industry recognizes three types of pools, including private–residential pools (sometimes called backyard pools), semi-public pools, and public swimming pools. The number of private pools significantly exceeds the number of semi-public and public swimming pools.

Semi-public pools are pools located at hotels, resorts, health clubs, condominiums, clubs, trailer parks, campgrounds, universities, and housing developments. They are usually intended for use by owners and invited guests. Public pools include recreational pools, competitive pools designed for swimming or diving (usually rectangle-shaped pools of varying size), and pools designed for special activities, such as water parks. In the case of public pools, many are designed to serve two functions: recreational and competitive. In addition, recreational pools often serve other functions. They are often used to teach SCUBA diving, provide a venue for water fitness classes, and to assist in hospital rehabilitation programs. In addition to the above, there are a number of small pools located on cruise ships that do not neatly fall into any of the previously referenced categories. Nevertheless, these pools must adhere to the strict codes, standards, and regulations of their land-based counterparts.

The number of fatal drownings in America is estimated to be approximately 4,000 per annum. Young children (aged one to four years old) or toddlers are significantly overrepresented in this number. The Centers for Disease Control reports that about 500 hundred young children drown in swimming pools ever year. This number makes drowning among toddlers and children the second leading cause of accidental death and swimming pools the major source of fatal drownings. It should be of no surprise that most swimming pool fatal drownings among children occur in private pools, followed by semi-public pools. Drowning also occurs at public pools, but these are relatively rare events because they are protected by lifeguards.

The CDC, American Red Cross, and the YMCA of the USA report that most fatal drownings are preventable. While estimates about the percentage of preventable drowning vary, the percentage of drowning that are preventable may exceed 90 percent. As mentioned earlier, about 40 percent of aquatic accidents are related to alcohol. Consequently, perhaps nearly half of all drownings are preventable by simply not drinking before engaging in an aquatic activity. In an unpublished study involving hot tub and spa accidents conducted by Fletemeyer, it was discovered that the major cause of drowning accidents is due to alcohol and drug abuse. A significant contributing cause of accidents in hot tubs or spas is bathing alone.

Key to the importance of swimming pool safety, whether at a private pool, semi-public pool, or public pool, is the concept known in the aquatic industry as *layers of protection*. This concept is founded on the belief that there are several safety measures that should be taken and that when combined, these measures contribute to the overall safety of pools and help to prevent a drowning event. This concept frequently appears in complaints alleging swimming pool negligence and consequently, its significance must be understood by attorneys, no matter what side of the law they practice.

Each of the swimming pool categories has some similar layers of protection in common but they also have layers that are unique to the category that they represent. For example, in the category of public pools, the primary layer of protection is the presence of certified lifeguards practicing the 10/20 rule. When considering layers of protection, it is important to note that some measures may be more important than others but when combined, they all play a part in the pool's safety. Ultimately and in theory, when all the layers are functioning as intended, they create a safe swimming pool environment that is relatively immune to a drowning event. However, when one or more are absent or are not functioning as intended or when alcohol and reckless or risk-taking behavior is involved, a drowning event or accident becomes inevitable.

Private pools are especially dangerous to children because pools and the water that they contain represent an attractive nuisance (see Figure 5.1). Consequently, private pools are magnets attracting children who are not supervised or do not have appropriate barriers.

There are several layers of protection when considering private residential pools. One of the most important is a barrier fence that surrounds the entire pool. In some cases, a side of a building can be substituted for a side of a fence but only if it does not have an access point, such as an open door or low-hung window that can be climbed through (more about this is provided later). Other important layers include a self-latching gate at least 48 inches high, a certified swimming pool cover that can cover a pool when not in use, a swimming pool alarm that will sound a warning if a child falls into the pool, and a safety line across the sides of the pool that defines where the shallow water becomes deeper. An additional layer includes door alarms and special locks that prevent opening by a child. Another measure that also may be regarded as a layer of protection is an automated external defibrillator (AED). More and more private pool owners have recognized the importance of an AED, and as their costs are lowering, they are purchasing them for their pools. In addition to the above and perhaps the most important layer of protection is responsible adult supervision during the time that a pool is being used. Adult supervision is not only important for children but for users of all ages.

In a semi-public pool, the layers of protection include many of the layers mentioned previously for private pools, including a barrier fences and a self-latching gate

FIGURE 5.1 Private residential swimming pool.

Swimming Pool and Spa Safety 77

FIGURE 5.2 Semi-public swimming pools.

(see Figure 5.2). There also three additional layers that are usually legislated into law by local and state by codes. These include the following:

- Safety equipment, including a shepherd's hook and a lifesaving buoy attached by a throw line
- A dedicated 911 telephone conspicuously located on the pool deck or attached to a wall inside the pool
- A floating buoy line that extends across the pool and delineates the deep end from the shallow end

If the pool is operated at night, another important layer of protection is lighting. Swimming pools must have sufficient lighting to, at the very least, see clearly the pool's bottom, including the drain.

In the category of public pools and water parks, the most important layer of protection is certified lifeguards that practice the 10/20 rule (see Figure 5.3). In recent years, many pools have added another important layer that provides the issuance of personal

FIGURE 5.3 Public swimming pools and water parks. These pools require certified lifeguards.

flotation devices (PFDs) to toddlers and to young children who cannot swim. A few pools have begun experimenting with underwater detection devices that are designed to alert a lifeguard or pool attendant if a victim is submerged underwater and is drowning. The verdict is still out regarding their effectiveness and their high cost is also an issue that relates to a cost–benefit analysis—does the benefit justify the high cost?

Public pools sometime serve as competition pools usually during early morning and late afternoon or evening hours. It is important to recognize that while mostly experienced swimmers are using the pool at this time, accidents and fatal drowning events still occur. Diving accidents from starting blocks and shallow water blackout from breath-holding exercises often occur at pools being used for competition. For this reason, pools involved in completion and training should remain supervised by lifeguards. There have been cases involving a coach who has assumed the role of lifeguard while coaching or instructing. In these cases, complaints alleging gross negligence have been filed. Coaches while actively coaching should not be assigned lifeguard duties. In addition, it is highly recommend that all swimming and diving coaches have CPR and emergency first-aid training. In addition to that, and as

Swimming Pool and Spa Safety

mentioned in the above, many pools serve additional functions. Pools involved in SCUBA training or water aerobics must always consider the safety of the participants.

HOT TUBS AND SPAS

A surprising number of hot tub and spa accidents occur in both private and public sittings. Drowning is a main hazard among hot tubs and spas. Since 1990, the CPSC reports that there have been more than 800 deaths in hot tubs and spas. About 20 percent of hot tub and spa fatal drownings involve children under the age of five. Also, a high percentage of drowning involved young adults and the elderly.

Since 1990, there have been forty-three incidents including twelve deaths in which a victim's hair was sucked into the suction fitting of a hot tub or spa. Hair entanglement occurs when a bather's hair becomes entangled in a drain cover as the water and hair is forcefully drawn by strong suction into the drain. In some incidents, children were playing breath-holding games that resulted in their body parts being trapped in the hot tub's or spa's drain.

Entrapment injuries are easily prevented by two drain outlets for each pump. This reduces the suction if one drain is inadvertently blocked by a body part. All spas, whether located at a private home or in a public setting, should be equipped with a clearly visible cut-off switch that functions to stop the pump and the suction that they are creating. In a notable case, the switch was hidden behind a boulder, making the switch difficult to locate thus delaying stopping the pump.

Hot tub and spa water temperature is another important source of injury. A temperature exceeding 110°F is considered dangerous. Elevated temperatures often cause drowsiness, which may lead to unconsciousness, resulting in drowning. In addition, hot water can cause heatstroke and death. For this reason, hot tub and spa temperatures must never exceed 104°F. Drug and alcohol abuse also results in drowning, especially among young adults.

Electrocution is another hot tub and spa hazard, especially among residential pools that have not been wired by professional, certified electrician. Electrical outlets should be provided with a cover and should be regularly inspected.

The CPSC recommends several safety precautions, including the following:

1. Always use a locked safety cover when the spa is not in use and keep children away from spas or hot tubs unless there is constant adult supervision.
2. Make sure the spa has the dual drains and drain covers required by current safety standards.
3. Regularly have a professional check your spa or hot tub and make sure it is in good, safe working condition and that drain covers are in place and not damaged or missing.
4. Know where the cut-off switch for your pump is so you can turn it off immediately. If using a hotel spa, locate the cut-off before entering.
5. Be aware that consuming alcohol or drugs while using a hot tub or spa could lead to drowning.
6. Keep the temperature of the water at or below 104°F.
7. Never bathe in a hot tub or spa alone.

TODDLERS

The Consumer Product Safety Commission (CPSC) and the Centers for Disease Control (CDC) reports the following information about drowning among toddlers and young children:

- Nearly 300 children younger than 5 drown in swimming pools and spas each year representing 75 percent of the 390 fatalities reported for children younger than 15.
- Children aged 1 to 3 represent 67 percent of the reported fatalities and 66 percent of reported injuries in pools and spas.
- More than 4,100 children younger than 5 suffer submersion injuries and require emergency room treatment; about half are seriously injured and are admitted to the hospital for further treatment.
- Males represent 80 percent of swimming pool drownings.
- African Americans are 5.5 times more like to drown in a pool than their white counterparts.
- The majority of drowning and submersion injuries involving victims younger than 5 occurs in pools owned by family, friends, and relatives.
- The majority of the estimated emergency department–treated submersion injuries reported as fatalities were associated with pools.
- Portable pools accounted for 10 percent of the total fatalities for children younger than 15.

There are many factors contributing to fatal drowning in swimming pools, but only one stands out when children are involved. The main reason why children drown in pools is the absence of responsible adult supervision (Fletemeyer, CDC, the American Red Cross, etc.). All too often, an adult becomes distracted by talking with a friend, making a phone call, texting, or reading a book while poolside.

There are two other reasons why toddlers and young children drown in swimming pools. Sometimes an adult will watch multiple children. The rule that should be followed is one adult for every child. The second reason is that the responsibility for watching a child is assigned to a sibling. In most cases, a young brother or sister should never be given this responsibility.

It has been well-established by drowning events that have been captured on video that drowning among children is usually relatively silent and seldom, if ever, is accompanied by an initial period of panic and struggle above the water. Instead of exhibiting these overt drowning behaviors, drowning among children is silent with few visible signs that an active drowning event is in progress. This is why it becomes critically important for an adult to practice "touch" supervision and to only watch one child at a time.

The second major reason why children drown involves pools constructed without child-proof barriers, pools with barriers that do not conform to recommended standards, or pools with barriers that have been compromised, that is, a gate that is not functioning properly. In the case of private swimming pools, it is the responsibility of the home owner to make sure that their pool has an appropriate barrier. In addition, the home owner must conduct periodic inspection and make repairs when needed.

RISK-TAKING BEHAVIOR

Another significant contributor to serious and sometimes fatal aquatic accident is risk-taking behavior, most commonly exhibited by teenage and young adult males. All too often while showing off to friends, a young male will perform a plunge or headfirst dive into shallow water and consequently will suffer a serious spinal injury that may result in quadriplegia and a life confined to a wheelchair. The cost of life care for a spinal injury patient is often astronomical (refer to the chapter in this book by St John), despite the fact that the victim's life expectancy is often significantly abbreviated as a result of the injury. Typically, in a spinal injury case involving shallow water, a plaintiff attorney claims negligence associated with appropriate warnings and water visibility. If the accident occurs at night, insufficient lighting may be a contributing factor, but ultimately, the aberrant behavior of the victim is to blame and supported by the courts and research conducted about this subject.

SECONDARY CONTRIBUTING FACTORS

There are a number of secondary factors contributing to injury and drowning in swimming pools. One of the objectives of this chapter is to identify and discuss some of the most important of these. Note that some of these contributing factors often occur in unison, thus escalating and exacerbating their seriousness. For example, the consumption of alcohol and a pool with a diving board represent two contributing factors and are often a recipe for a serious swimming pool accident.

Swimming Pool Drains and Entrapment Injuries

The Virginia Graeme Baker Pool and Spa Safety Act was passed by Congress in 2007. This act regulates drain safety, making anti-entrapment drains a legal requirement. Consequently, entrapment accidents causing drowning events seldom occur in pools located in the United States. However, some unregulated private pools have not been retrofitted with anti-entrapment drains. These pools increase the likelihood of an entrapment injury or death, especially among children with long hair. It is important to note that the Virginia Graeme Baker Act does not require residential pools to have anti-entrapment drains. Consequently, entrapment injuries and deaths are significantly more likely to occur in backyard swimming pools than in regulated semi-public and public pools.

All pool water outlet covers should be inspected regularly when the pool is in operation. If any of the water outlet covers are loose or missing, the pool should be closed until the cover is repaired or replaced. Note, loose or missing outlet covers have caused fatalities and serious injuries.

When inspecting an outlet covers, caution must be taken to ensure staff safety. The inspection procedure should include shutting down the filter systems and ensuring that

- There is no suction in the system.
- The system is locked-down or supervised to ensure that it is not turned on during the inspection.

- A diagram of the pool depicting the outlet cover will guide the inspection. Each outlet cover should be assigned a number to help the accurate recording of inspection results.
- A second person should be present as an emergency back-up during the inspection.
- The inspection results are recorded and a remedial action is required and completed.

It is important to note that in the absence of national legislation making anti-entrapment drains necessary, many pools located in other countries have not been retrofitted with anti-entrapment drains. At these pools, serious accidents are waiting to happen.

Swimming Pool Lighting

Many pools, especially those classified as semi-public pools, remain open at night. It is important for a pool operated at night to have sufficient lighting to clearly see the bottom of the pool at its deepest part; usually this is where the drain is located in the deep end. In addition, there must be enough lighting to see the pool's depth markers that are usually located on the pool's coping. While lighting standards vary from state to state, ANASI swimming pool standards recommend that there is sufficient light to illuminate the entire pool, including the deck and surrounding area. Usually this level illumination is between three and six lumens or foot candles of lighting and is often stated in state adopted codes. An unpublished study conducted by Fletemeyer found that a significant number of pools failed to have sufficient lighting for either one of two reasons (see Figure 5.4). Either their pool lighting was not properly engineered or lighting was not being properly maintained. At several pools investigated by Fletemeyer, lightbulbs had burned out and had not been replaced.

FIGURE 5.4 An insufficiently illuminated pool.

Swimming Pool Water Clarity

Poor water clarity is often a significant contributing factor to a fatal drowning event by preventing a caregiver or a lifeguard having clear vision of the water column and the pool bottom. A particular case that stands out that relates to this standard involved a Massachusetts public swimming pool where a young man drowned after the pool's deep end had been closed for swimming due to cloudy water. Swimming was allowed to continue but not in the deep end of the pool. The victim's body was discovered the following day by a lifeguard while vacuuming the pool bottom. He felt the vacuum bump into something—the victim's body. This is an illustration of gross negligence resulting from cloudy water and the decision not to close the pool.

Poor water quality significantly impacts seeing the latent stages of a drowning event and is often responsible for a delay when performing a rescue within the time parameter that permits the opportunity for successful resuscitation. As a rule of thumb, the deepest part of the pool (the drain located at the deep end) should be seen while standing anywhere on the pool deck. If the pool bottom cannot be seen, the pool should be immediately closed.

A more exact method of determining water quality involves the use of a Secchi disc (see Figure 5.5). A Secchi disk is a black disk that is 150 millimeters in diameter on a white background, is located on the bottom of the pool at its deepest point, and is clearly visible from any point on the deck 9 meters away from the disk.

The swimming pool water clarity (sometimes called visibility or turbidity) must be regularly monitored. In most cases, management of semi-public and public pools require that staff check the pool at least twice a day. Various factors contribute to a pool's water quality, including the following:

- An unbalanced water chemistry (see section on water chemistry)
- An increase in the pool's bathing load
- Meteorological conditions including sunlight, temperature, and rainfall

FIGURE 5.5 Photograph of author holding a Secchi disk used to test water clarity.

Experiments conducted by Fletemeyer suggests that under normal conditions, it takes two or three days (48 to 72 hours) for a pool to transition from a clear to a cloudy state. However, in unusual circumstances, the change can occur much more rapidly. Other experiments by Fletemeyer suggest that water becomes cloudy, making the bottom difficult to see, when the turbidly value reaches or exceeds 20 NTUs.

Swimming Pool Decks

An improperly maintained swimming pool deck represents a hazard that could lead to a serious accident. This hazard must be addressed by swimming pool owners or their delegated management and staff. Unless alcohol, running, or horseplay is involved, a slip-and-fall injury is preventable and consequently, negligence is usually to blame.

Poor or improper maintenance is usually the reason for a slippery deck. All pools, no matter to which of the three categories they belong (residential, semi-public, or commercial), must be inspected daily for water puddles, algae, bacteria growth, and oily containment build up. Deck areas located in and around outdoor showers and shaded areas require special attention because these areas often create an environment promoting algae growth. In shower areas, residual sun tan oil, soap residue, and hair oil is often responsible for creating a slippery deck surface. Cruise ships are especially prone to poolside slip-and-fall accidents because the deck is seldom stable due to the rocking and rolling action of the ship. Also, a fiberglass deck often contributes to a slippery surface. Various industry-accepted methods are prescribed to keep decks slip-free, including the following:

- Decks should be regularly hosed with high-pressure water.
- Detergent should be used in areas prone to sun tan, sunscreen, and hair oil build up, for example, shower areas and locker room entrances.
- In bad cases, chlorine should be used to kill and eliminate algae growth. Typically a solution should be one part chlorine to ten parts water (= 10 percent solution).
- In the worst cases, a weak solution of muriatic acid can used but extreme caution should be taken when applying. Plastic gloves, boots, and a gas mask should be used. An acid wash should never be conducted when bathers are present in the pool.
- Water puddles can be temporarily addressed by squeegeeing the pool deck. However, this must be regarded as a temporary solution. Low depressions or cracks in the pool deck should be filled with "leveling" cement to eliminate this problem.

When pool decks become slippery, special caution "Slippery When Wet" cones should be placed in the areas.

A swimming pool log should be maintained listing the measures taken each day to eliminate slippery decks and pool surfaces. A maintenance log becomes useful in defending a slip-and-fall lawsuit. Additionally, more and more pools are using surveillance video cameras. In the event of a slip-and-fall injury, surveillance video can be analyzed to determine the accuracy of a victim's description of his or her injury.

Swimming Pool and Spa Safety

Diving Boards

At one time, diving boards were common at most pools and were often the source of serious injury and frequent litigation claims. Thanks to the advice and warnings from the insurance industry, diving boards have been removed from most pools.

Minimum standards for safe entries off a diving board or platform are provided by the Federation International de Natation Amateur (FINA). One of FINA's standards is that when diving from a 1 meter platform, the water must be at least 11.5 feet deep. Many older pools, especially those located in private homes, often do not have diving boards that comply with FINA standards. These diving boards should be removed from the premises.

Barrier Fences

As stated earlier, barrier fences represent an important layer of protection at all three categories of pools including private, semi-public, and public. While barriers sometime differ in dimension, ANSI and CPSC standards should apply. It is important to note that barrier fences are not childproof but often give parents and caregivers additional time to find a child if discovered missing. The following are recommendations by the CPSC about effective swimming pool barriers:

1. The top of the barrier should be at least 48 inches above the ground surface measured on the side of the barrier that faces away from the swimming pool.
2. The maximum vertical clearance between the surface and the bottom of the barrier should be 4 inches measured on the side of the barrier that faces away from the swimming pool. In the case of a non-solid surface, grass, or pebbles, the distance should be reduced to 2 inches, and 1 inch for removable mesh fences.
3. Where the top of the pool structure is above grade or surface, such as an above-ground swimming pool, the barrier may be at ground level, such as the pool structure, or mounted on top of the pool structure. Where the barrier is mounted on top of the pool structure, the maximum vertical clearance between the top of the pool structure and the bottom of the barrier should be 4 inches.
4. Openings in the barrier should not allow passage of a 4-inch diameter sphere (see Figure 5.6).
5. Solid barriers that do not have openings, such as masonry or stone walls, should not contain indentations or protrusions except for normal construction tolerances and tooled masonry joints.
6. Where the barrier is composed of horizontal and vertical members and the distance between the bottom and top horizontal is less than 45 inches, the horizontal members should be located on the swimming pool side of the fence.
7. Spacing between the vertical members should not exceed 1¼ inches in width. Where there are decorative cutouts, spacing within the cutouts should not exceed 1¾ inches in width.

FIGURE 5.6 Photo of author holding 4-inch sphere next to fence.

8. Maximum mesh size of chain link fences should not exceed 1¼-inch squares unless the fence is provided with slats fastened at the top or the bottom which reduce the openings to no more than 1¾ inches.
9. Where the barrier is composed of diagonal members, such as a lattice fence, the maximum opening formed by the diagonal should be no more that 1¾ inches.
10. Access gates to the pool should be equipped with locking devices. Pedestrian access gates should open outward, away from the pool, and should be self-closing and have a self-latching device. Gates should have a self-latching device where the release mechanism of the self-latching device is located less than 54 inches from the bottom of the gate.
    a. The release mechanism should be located on the pool side of the gate at least 3 inches below the top of the gate.
    b. The gate and barrier should have no opening greater than ½ inches within 18 inches of the release mechanism.
11. Where the wall of a dwelling serves as part of the barrier, one of the following should apply:
    a. All doors with direct access to the pool through that wall should be equipped with an alarm that produces an audible warning when the door and its screen, if present, are opened. Alarms should meet the requirements of UL 2017 General-Purpose Signaling Devices and Systems, Section 77.
    b. The pool should be equipped with a power safety cover which complies with ASTM F1346-91.—"Standard Performance for Safety Covers and Labeling Requirements for All Covers for Swimming Pools, Spas and Hot Tubs."
    c. Other means of protection, such as self-closing doors with self-latching devices, are acceptable so long as the degree of protection afforded is not less than the protection afforded by (a) or (b) described above.

Swimming Pool and Spa Safety 87

12. Where an above-ground pool structure is used as a barrier or where the barrier is mounted on top of the pool structure and the means of access is a ladder or steps, then
    a. The ladder to the pool or steps should be capable of being secured, locked, or removed to prevent access.
    b. The ladder or steps should be surrounded by a barrier. When the ladder or steps are secured, locked, or removed, any opening created should not allow the passage of a 4-inch sphere.

Gates

Swimming pool barriers usually have a gate allowing access and egress. Gates should open out from the pool and should be self-closing and self-latching. If the gate works properly and for some reason is not properly latching, a child pushing on the gate to enter the pool area will at least close the gate and engage the latching mechanism.

When the release mechanism of the self-latching gates is less than 54 inches from the bottom of the gate, the release mechanism for the gate should be at least 3 inches below the top of the gate on the side facing the pool (see Figure 5.7). Placing the

FIGURE 5.7 Photo of latching mechanism.

release mechanism at this height prevents a young child from reaching over the top of the gate and releasing the latch.

The gate should have no opening greater than a half-inch within 18 inches of the latch release mechanism. Gates sometimes malfunction. Children have a habit of swinging on them. This behavior sometimes causes the gate to remain open. Consequently, because gates are so important in keeping unsupervised children out of the pool, they should be regularly checked and promptly repaired if not working properly.

Waterslides

Waterslides are often the source of an accident and should be maintained according to the instructions supplied by the manufacturer. Rules should be established that minimize the risk of collision or injury within the slide or the landing pool/flume at the bottom of the slide. Note that sliders often intentionally slow down their speed and this can create a safety risk that may be a contributing factor to a waterslide injury. Also when several bathers slide at once, this can be dangerous, resulting in a serious spinal or head injury.

Waterslides, especially those located at water theme parks, should be diligently supervised with rules requiring their safe use.

Swimming Pool Design and Engineering

The majority of swimming pools have deep ends usually exceeding the height of an adult. Pools with deep ends usually have a slope transition from shallow water to deeper water. However, a few usually older pools have sudden drops-offs with no gradual transition.

The transition between shallow and deep water is often the source of a drowning event. A non-swimmer or poor swimmer will gravitate from the shallow end into the deeper end often loose footing with the bottom and suddenly and unexpectedly be in water that is overhead. For this reason, most states require that pools have a permanent float line extending across the pool that defines the shallow end from the deep end. In addition, some states also require a painted line on the bottom of the pool also defining the transitioning to deeper water.

When fatal drownings occur in pools that do not have a painted bottom line and a float line, this is often the subject of a complaint and listed as a contributing or proximate cause.

Signage

Signs primarily at semi-public and public pools serve two and sometimes not always mutually exclusive functions.

- Signs serve to warn about a hazard that someone may not know existed.
- Signs serve as reminders about hazards and conditions that may result in injury or death if not followed.

In addition, signs provide the following information to a swimming pool user:

- Information about supervision requirements.
- Information about the suggested rules for safe use of the facility.
- Warns users of hazards and ways to avoid these hazards.

Swimming Pool and Spa Safety

Signs should always be posted in conspicuous locations in the pool area. In some cases, signs become hidden by overgrown vegetation or placed in a location that cannot be easily seen. In some cases they are located only a few feet above the ground. Consequently, a lounge chair with its back elevated or a person standing poolside may prevent the sign from being observed. Vegetation is often responsible for hiding signs and should be regularly maintained. In addition, when signs start to deteriorate and become faded or rusted, they lose their effectiveness. They also lose their effectiveness when too many signs are placed together. Sometimes, this condition is called "sign pollution."

In America, the ANSI recommendations for warning signs should be adopted for use in swimming pools. The recommendations by ANSI require warning signs to be represented in a certain color, shape, and size. Any deviations from ANSI recommendations could be the subject in a complaint in response to an accident or fatal drowning.

In addition to warning about hazards, it become incumbent to establish the following rules that are related to admission policies, including:

- Minimum age and requirements for supervision of children (see Figure 5.8).
- Notification of medical conditions that may affect bather safety (i.e., seizure disorder, potential fecal contamination).
- Whether non-guests/tenants are allowed access to the pool.
- The time of day that the pool is open or closed for swimming.

In additions, signs should be posted to establish rules against other activities that may be considered injurious to bathers, including:

- No breath-holding games or activities.
- No playing with hard balls and aerial disks (Frisbees).
- No running or engaging in horseplay.

FIGURE 5.8 Photo of sign requiring children to be supervised.

- No bringing objects made of glass into the pool.
- No drinking alcoholic beverages.

Most semi-public and public pools also rely on warning logos. Logos are especially useful in pools where there are foreign visitors who are not able to read and understand signs written in English. Typically logo signs display a representation of the activity meant to be prohibited (i.e., "No Diving") and a red line running diagonally through the activity meant to be prevented. In swimming pools located in areas where there is more than one clearly defined ethnic group, signs are sometime presented in more than one language. For example, in South Florida, swimming pool warning signs are sometimes written in English and Spanish.

In summary, signs should be clearly written and unambiguous. They should be located within the vision field of the guest upon entering the swimming pool and their condition should be frequently checked and should be replaced the moment they begin to deteriorate.

Depth Markers

All states require that semi-public and public pools have depth markers strategically placed around the edge or coping of the swimming pool. Often there is also a requirement that the depth makers be accompanied by "No Diving" logos where the pools depth is too shallow to permit plunge or head-first diving (see Figure 5.9). A study conducted by Fletemeyer found that most semi-public private and public swimming pools display both depth markers and "No Diving" logos and were in compliance with national swimming pool standards. A survey of private pools found that only a relatively small percentage had depth markers and an even smaller percentage of pools had "No Diving" logos.

However, despite the presence of appropriate depth markers and logo warnings at semi-public and public pools, many pools that were evaluated did not have sufficient light that permitted seeing the depth markers. Consequently, this may be one of the reasons why shallow water diving accidents occur more often during nighttime swimming.

Hot Tubs and Spa Signs and Safety

Safety rules for whirlpools commonly include some of or all of the following examples:

- Check for safe temperature. The temperature should never exceed 104°F.
- Enter and exit slowly. Headache or dizziness are signs to leave the water immediately.
- Never use the spa alone.
- Limit length of use to 10–15 minutes at a time. Note: A clock should be clearly visible from the spa.
- Children under twelve years of age should be supervised by an adult at all times. Children under five years of age should not be allowed in the spa (see Figure 5.10).

Swimming Pool and Spa Safety

FIGURE 5.9 Photos of "No Diving" warning logos. (*Continued*)

FIGURE 5.9 (CONTINUED) Photo of "No Diving" warning logos.

FIGURE 5.10 "No Children Allowed in the Spa" notice.

- Pregnant women should only use a spa or hot tub with the approval of their doctor.
- Persons suffering from heart disease, diabetes, and high or low blood pressure should consult their doctor prior to use.
- Do not use the spa or hot tub while under the influence of alcohol or drugs.

Swimming Pool Bathing Load

Various formulas are used to calculate a swimming pool's bathing load, including the dimensions of the pool's foot print, water depth, and bathroom capacity. In the majority of cases, semi-public and public pools must use signage to state the bathing load. People in the general use area (such as the areas set aside for loungers or for watching swimmers)

Swimming Pool and Spa Safety 93

are not considered bathers. However, if people cross over from the general area onto the deck or into the pool, then they become part of the pool's bathing load.

Exceeding a pool's bathing load may be a contributing factor in a drowning event because too many bathers prevent carefully monitoring swimmers, especially non-swimmers or fair swimmers. In addition, exceeding the bathing load is sometimes responsible for making the water cloudy and impacting the pool's water chemistry. For these reasons, an overcrowded pool may become a contributing factor in a negligence suit.

Bathing loads are often exceeded in smaller, privately owned swimming pools. Birthday parties and other social events should be carefully monitored to prevent the bathing load from being exceeded.

Water Chemistry

Contrary to popular belief, chlorine doe not kill all germs instantly. There are many germs that are responsible for causing serious illness that are resistant to chlorinated water. Consequently, ingesting only a small amount of water containing infectious germs can make a bather ill.

Recreational water illnesses (RWIs) are caused by germs spread by swallowing, breathing in mists, or having contact with contaminated water in swimming pools, water play areas, and spas. Spas are particular dangerous because of elevated water temperatures, which provide the perfect environment for germ reproduction.

RWIs include a wide variety of infections including gastrointestinal, skin, ear, respiratory, eye, neurologic, and wound infections. The most common RWI is diarrhea. Diarrheal illnesses are often caused by *Crypto* (*Cryptosporidium*).

Studies of *Crypto* and other infections and swimming pool illnesses supported and funded by the National Swimming Pool and Spa Institute suggests that waterborne illnesses are increasing, especially illnesses caused by *Crypto*.

Recently studies focusing on *Crypto* report that this organism often remains alive for days even in pools that are well-maintained. From 2004 to 2008, reported *Crypto* cases increased over 200 percent (from 3,411 cases in 2004 to 10,500 cases in 2008).

Although *Crypto* demonstrates a tolerance to chorine, most illness-causing germs do not. Consequently, chorine and pH levels must be properly maintained. However, it is important to note that a study conducted by the CDC (2008) reported that 1 in 8 public pool inspections resulted in pools being closed due to serous code violations such as improper chlorine levels.

As mentioned earlier, the objective of this chapter is not to discuss pool and pool chemistry in detail but to provide an overview. Having the appropriate and maintaining the recommended water chemistry is important for the reasons stated previously but it is also important for the following two reasons:

- Having inappropriate or unbalanced water chemistry may result in cloudy water, resulting in minimal visibility. As a rule of thumb, swimming pool water must always be maintained at a level allowing the deepest part of the pool to be seen while standing on the pool deck. Experiments conducted by Fletemeyer found that cloudy water begins when turbidity values exceed 20 NTUs.

- Having inappropriate water chemistry may impact a bather's overall recreational experience. A highly chlorinated pool with an unbalanced pH may cause a swimmer's eyes to burn.

For primarily these first two reasons, pool chemistry must regularly monitored. In some states, semi- public and public pools must be checked twice daily to make certain the water pH and chlorine values are within an acceptable range. These twice-daily checks should be recorded in a swimming pool logbook for future reference and verification. It is important to note that sometimes during heavy rains or periods of intense sunshine, chemical values can quickly change. Consequently, additional checks may be necessary. A heavy bathing load (defined as the total number of bathers in a pool at one time), rain, and sunshine may have a significant impact on the swimming pool's water chemistry and will dramatically decrease a pool's water clarity.

Pool chemistry for the three types of pools discussed in this chapter remains the same. Specifically, the free available chlorine residue should be maintained at a minimum of 1.0 ppm for any swimming pool with an operating temperature of not more than 30°C and 2.0 ppm for any pool temperature of more than 30°C. The pH value must be maintained at not less than 7.0 ppm and not more than 7.6 ppm.

The Centers for Disease Control has published a standard for the disinfection of water in a whirlpool, hot tub, or spa that is significantly higher than the values stated previously for swimming pools. A minimum free available chlorine (FAC) of 4.0–5.0 p.m. should be maintained in all whirlpools, hot tubs, and spas.

Safety Equipment

All three categories of swimming pools (private, semi-public, and public) should have strategically located and properly maintained safety equipment that can be used to assist during an active drowning event. Usually this includes a shepherd's hook that extends at least to the middle of the pool's bottom if not beyond, a lifesaving throw ring (buoy) with an attached 3/8-inch polypropylene line, and a dedicated emergency 911 number (see Figure 5.11). In larger pools, there may be a need to have several of these safety items.

The requirement for pools to have the above reference safety items varies from state to state. In the majority of cases, semi-public pools are required by code or law to have this equipment and to have it properly maintained. There is not a similar requirement for private pools to have this equipment.

Lifeguard Protection

In a majority of cases, semi-public pools are not required to have lifeguards or attendants. However, there are a few notable exceptions where resorts or hotels have voluntarily provided lifeguard protection to their guests. Only public pools including water parks are required to have lifeguards. Several organizations train swimming pool and water park lifeguards, including the America Red Cross, the YMCA of the USA, Ellis and Associates, and the Boy Scouts of America. Of these organizations, the America Red Cross trains the largest number of lifeguards. Most certifications remain current for two years and then must be renewed. Note that in a drowning

Swimming Pool and Spa Safety 95

FIGURE 5.11 Photo of shepherd's hook and buoy.

event involving a lifeguard having an expired certification card, it may be legitimately argued that the pool was unguarded. It is the duty of both the supervisor and lifeguards to make sure cards are current.

While the time recommended to identify a victim in trouble and the response time to effect a rescue varies among different lifeguard organizations, in most cases, lifeguards are trained to rely on scanning methods and to identify an actively drowning event within 10 seconds. Once observed, they are then required to respond within a 20-second window. Sometimes this is called the "10-20" rule. Failure of a lifeguard to identify and respond to a drowning event expeditiously often leads to a lawsuit claiming gross negligence and a neglect of duty.

While lifeguards significantly increase the safety of a swimming pool, they do not immunize a pool from a drowning. There are numerous cases involving drownings at guarded pools. During a recent paper presented by the Redwood Group, it was reported that about 60 percent of the time, a member of the public observes a drowning event before a lifeguard. Fletemeyer reported that despite training, many lifeguards, especially young ones, still do not know how to accurately recognize a drowning victim. In a study by Fletemeyer (2014), individuals with lifeguard training were shown a movie scene that inaccurately portrayed a drowning event. About 20 percent of the lifeguards believed that this was an accurate representation of a drowning.

Consequently, several cases have been settled for large amounts involving lifeguards who failed to observe a drowning victim and act with a timely response. The following represents negligence factors involving lifeguards:

- Lifeguards without active certifications.
- Lifeguards who are distracted, for example, using a cell phone or engaging in a social conversation.
- Lifeguards assigned to other tasks besides watching the water, for example, cleaning the pool deck, getting towels for patrons, or collecting user fees.
- An insufficient number of lifeguards. One lifeguard is never enough, even at a small pool. Lifeguards must be allowed time to take breaks, eat lunch, and use the restroom. At large pools, several lifeguards are needed, allowing for well-defined areas or "zones" of coverage with overlapping vision fields.
- Lifeguards should not be required to watch toddlers. In some pools lifeguards are required to watch children in the absence of a parent or responsible adult. Never should this practice be allowed at a swimming pool.

Another contributing factor that sometimes relates to negligence claims is recurrent training. Pool management must carefully plan an ongoing, recurrent training program, allowing lifeguards to practice, rehearse, and hone their lifeguarding skills.

Overseas Swimming Pool Safety

American pools are usually strictly regulated by state and local codes. Pools located in other countries, especially pools located in what are sometimes called third-world countries are not similarly regulated and consequently are often dangerous for a variety of obvious and not-so-obvious reasons. In a study conducted by Fletemeyer of eighty-six pools located in seven countries, nearly every pool had a significant safety violation requiring immediate attention. A surprising and alarming finding reported in this study was that large hotel corporations based in the United States often do not have their pools located overseas adhere to the same American safety standards. Americans visiting American-based hotels overseas expect a safe and hazard-free swimming pool environment.

Other (Special Events, Parties, Shallow Water Blackout, Sexual Predators, and Emergency Closures)

There are several other factors that may contribute to negligence and a pool's exposure to a lawsuit. Some pools permit special functions, including barbecues, birthday parties, swimming meets, and movie nights. These special functions often require adding additional lifeguard staff and planning by pool management. During special events, management and staff must be careful that the event does not exceed the bathing load. In the case of movie night, this special event should be discontinued because pool lighting is often turned off to see the poolside movie. Consequently, guards are no longer able to carefully monitor the pool and observe the initial stages of drowning behavior.

Lifeguards are also trained to be vigilant for other forms of behavior that might lead to a drowning. Breath-holding competitions and games may lead to a life-threatening condition called *shallow water blackout*. Lifeguards must stop any and all activities involving breath-holding among patrons.

Lifeguards must attempt to profile patrons for signs of intoxication. No one suspected of drinking should ever be allowed in a swimming pool and no patron should ever be allowed to drink alcohol while in a pool or while poolside. An especially dangerous practice occurs in semi-public pools (hotels and resorts) that allow drinking inside a pool. This same inappropriate condition often occurs on cruise ships. Drinking in pools or on pool decks is an accident waiting to happen.

Pools often attract sexual predators. Lifeguards must always be vigilant for individuals with cameras taking photographs of young children or acting inappropriately. Locker rooms and pool bathrooms should be regularly checked. While profiling seems to be politically insensitive, lifeguards must be aware that middle-aged white males typically fit the profile of a sexual predator.

Sick people should never be allowed in the pool. Lifeguards should always be watchful for anyone with open sores or visible wounds. Pool policies must be established dealing with this potential threat.

Non-swimmers must also be responded to proactively. When a child is identified as a non-swimmer or even a fair swimmer, many pools require the use of PFDs. Some pools issue PFDs to non-swimmers.

Electrical storms represent a challenge at guarded pools. Once an electrical storm is identified, a swimming pool, including the pool deck, should be expeditiously cleared of patrons. Some pools rely on lightning detection devices to assist with the identification of nearby electrical storms.

There are more than eight million pools in America. Swimming pools are classified as either being in-the-ground pools or above-ground pools. Most above-ground pools are private pools located at private homes.

REFERENCE

CDC. (2008). Violations identified from routine swimming pool inspections—Selected states and countries, United States, 2008. *Morb Wkly Rep.* 59:582–587.

6 Headfirst Recreational Aquatic Accidents

Tim O'Brien

CONTENTS

A Brief History 100
A Comparison 102
A Diving Injury and a Headfirst Recreational Aquatic Accident 103
   So What Happened? 108
What Is a Spinal Cord Injury (SCI)? 108
   The Statistics 110
   The Facts 112
105 Diving Accidents Involving Aboveground Pools 114
211 Diving Accidents Involving Inground Pools 116
92 Diving Accidents Involving a Springboard/Jumpboard 118
32 Diving Accidents from Swimming Starting Blocks 121
161 Diving Accidents in the Natural Environment 123
The Number-One Cause of Headfirst Recreational Aquatic Accidents:
Shallow Water 125
How Can We Reduce Diving Accidents? 127
   Pool Design and Recommended Changes for Safer Pools 128
      Spoon, Hopper, and Vinyl Bottom Pools 128
      Markings on Pool Bottoms 129
      Starting Blocks 129
      Water Slides 129
      Depth Markers 129
      Florescent, Solar or Neon Nighttime Depth Markers 130
      Locations of Pool 130
      Warning Signs 130
      Lighting 130
      Barriers 130
      Basic Water Depth Awareness 130
      Elimination of Safety Ledges 131
      Marked Edges or Seats 131
      Recessed Steps, Ladders, Anchors, and Receptacles 131
      Lifeline 131
      Color-Coded Pool Edges 131
      Improved Instructional Signs for All Pool Types and Locations with Boards ... 131

Better Pool Management and Operation 132
    Better Supervision 132
    Industry Standards 133
    State and Local Laws and Regulations... 133
Operation and Management by Residential Pool Owners 133
Operation and Management by Hotel/Motel, Condo/Apartment, and Public Pools (in Addition to the Guidelines Above) 134
    Recommendations to the Insurance and Medical Industry 134
A Note for Parents 135
In Closing 136
Reference 136

As you will come to understand in the following pages, and as those in the sport of diving know, thousands of senseless and preventable aquatic accidents occur each year and are classified as *diving accidents*. The worldwide stigma to the Olympic sport of diving is unwarranted. The reality is that these tragedies typically involve self-taught, recreational swimmers going headfirst into various *shallow* aquatic environments. Most often, they are males. They are not "divers," nor are these "diving accidents," nor are they in any way related to the sport of diving. They are typically recreational swimmers-turned-untrained divers entering the water headfirst with catastrophic results.

A BRIEF HISTORY

As a young boy growing up in Columbus, Ohio, and the son of an eight-time U.S. Olympic diving coach, time in our household was measured in four-year increments and focused on the next Olympic Games. My earliest idols were the greats of springboard and platform diving, some of whom were the Ohio State divers of that era and others who were great champions from places like Mexico, Italy, Russia, and East Germany. Later, as an All-American diver at the University of Miami, I was fortunate enough to compete during what is regarded today as the "greatest generation" of American divers that have ever been produced. There was so much talent in the United States in the 1980s that many of the divers who did not make our Olympic Team due to the limit of two Olympic qualifiers per country would have had a chance at medals or to make the finals in the Games themselves. The depth of American diving talent was astonishing. I spent my entire youth training, traveling, and competing against the finest divers in the world and was blessed to have witnessed the greatest champions of the modern era. Years later, as one of the U.S. Olympic diving coaches at the 2000 Games in Sydney, Australia, I got to coach with my dearest friends and to see my diver, Michelle Davison, compete in the women's three-meter finals. Today, many of the coaches, athletes, volunteers, and judges in the United States as well as those in the international diving community remain my second family. The athletes I competed with, the coaches I had the honor to work alongside, the joy of competition, the camaraderie, and the lessons taught through sport have been a true gift and are part of some of the best days of my life.

That arena, the beauty and excitement of competitive diving, could not be any further from the sheer terror, grief, sadness, and helplessness that exists in a much different world; providing expert witness testimony, advice, and opinions in senseless accidents where someone's life, and family, has been irreparably damaged. But this world exists. Needless to say, the competitive diving community throughout the world is sensitive to the unfairness and inaccuracy of labeling the accidents that occur involving untrained swimmers entering the water headfirst, whether by the fault of the participant, the facilities, or those managing them, as "diving accidents." But certainly, the negative impact these accidents have on the sport itself in no way compares to the people, the injuries, and the families that are left to pick up the pieces.

As you can imagine, this unfair classification has led to increased insurance costs, dwindling pools, facilities, equipment, and a general, but completely false, impression by the public that the *sport* of diving is unsafe. As someone who has spent more than five decades in the sport, I can attest to the fact that accidents in the sport of diving are rare. From time to time, a trained competitive diver may hit their feet or hands on the springboard, slip off the board into the pool, or smack on a new dive, but the catastrophic types of injuries that we will discuss in the following pages simply don't occur. Compression of the cervical spine does not occur in the sport of diving because well-trained divers participate in facilities worldwide with adequate water depth. Most pools made for competitive diving are 12–18 feet in depth, providing ample room for a trained diver to maneuver. Injuries to competitive divers at the world class and Olympic level typically involve strains, sprains, or "overuse syndrome" of the wrists, shoulders, back, ankles, and neck. The impact sustained while entering the water headfirst from the 10-meter platform can cause competitive divers to protect their thumbs and wrists with tape or small braces, or take extended time away from that event to heal. But, as stated, compression of the cervical spine injuries do not occur. Now, it must be said that even at a depth of 18 feet, a six-foot-tall person entering the water headfirst from a one-meter diving board or doing a running dive from the deck could strike the bottom with sufficient velocity to generate a compressive force great enough to injure the cervical spine if that person were to strike the bottom in the "head-ducked" position. The dangers that exist in the sport of diving are with equipment, the springboard or platform, that the diver takes off from and the surface of the water itself—not its depth. The recreational diver is at greatest risk because of a lack of training, the overestimation of his or her skills, the underestimation of the risk, and the shallow water that they have not been taught to avoid at all costs. Classifying headfirst recreational aquatic accidents as "diving accidents" or attributing them to the sport in any way is akin to categorizing car accidents to NASCAR.

As a parent myself, I urge all parents to enroll their children in basic diving lessons with a United States diving-certified coach who can teach children safety lessons and awareness that they can put into practice. These skills acquired could protect them when they visit a friend's backyard pool for the first time, feel pressured by their peers, consider diving into a body of water they are unfamiliar with, or try to show off to their buddies. You can visit USA Diving's website at www.usadiving.org to find a certified, qualified coach in your area. A good mantra for parents and swimmers to remember and convey to others is the saying "feet first, first" when entering any pool or body of water.

As an expert witness, I am associated with individuals who have sustained horrifying spinal cord, paralysis, or life-threatening injuries that are mostly preventable. In a word, the experience is heartbreaking. I have looked at photos and watched video replays of accidents with tears in my eyes because I know that these accidents are unnecessary. I cringe when I see them or have to watch a moment of impact over and over again. The statistics and data are out there, and prevention really lies in just simple common sense. But in many cases, with unforeseen circumstances, peer pressure, alcohol and drugs, unsafe facilities and pools, bad or worn-out equipment, poor supervision, or just plain stupidity, accidents in the aquatic environment happen. It is my hope, more than anything, that my involvement and my voice will lead to more awareness and prevention. I do not want anyone's loved ones to have to go through such a tragedy. I hope the information presented here will further educate the public, pool builders, designers, facility operators, and regulatory agencies, and that people, especially youngsters, will heed the warnings about shallow water, unknown bodies of water, unfamiliar pools and environments. Remember—"feet first, first."

A COMPARISON

In a notable publication, *Diving Injuries* (2000), written by well-known experts on the subject of diving injuries, Dr. M. Alexander Gabrielsen falsely writes in the preface that

> Diving, without question, is the most exciting and challenging activity associated with water, yet diving produces more quadriplegics than all other sports combined. Today, there are an estimated 19,000 people sitting in wheelchairs as a result of breaking their necks when diving.

While the data from this book is dissected throughout this chapter, it must be stated that this is an entirely untrue statement based on his own research, data, and findings. In my opinion, the use of the word *diving* gives people the false impression that these accidents are related to the sport and to trained athletes when they are not. The irresponsible use of the word *diving* or any connection to the *sport of diving* is false and totally misleading to the public, especially parents. The sport of diving does not produce quadriplegics, as Gabrielsen states. Once again, spinal cord injury accidents, like the hundreds he himself studied, are a product of untrained, novice, recreational swimmers who decide to dive into bodies of water headfirst. The responsibility often lies with the "victim" or the many other entities, facilities, officials, designers, owners, and lack of proper supervision of the participants, but in no way are they associated with the sport of diving. Even the experts must be careful with their use of the word *diving* and what it might falsely imply to the public.

Gabrielsen goes on later in the preface to state correctly:

> Competitive divers sustain very few injuries. To date, there is no record of any competitive diver receiving a neck fracture by striking the bottom of the pool that met the specifications promulgated by the various ruling bodies governing competitive diving. As the data in this book will reveal, the recreational diver is at greatest risk.

Nonetheless, it is important to be responsible in the use of language in describing these accidents so that parents and the public understand who is at risk and to be clear on how these accidents occur.

A DIVING INJURY AND A HEADFIRST RECREATIONAL AQUATIC ACCIDENT

In August 1988, at the age of 25, I sat in the Jamsil Swimming and Diving venue during the preliminaries of the men's three-meter diving event at the 1988 Seoul Olympic Games. My father stood on the pool deck just below as coach of the U.S. Diving Team in his sixth Olympic Games in that role. His diver was performing well during the event; a long, grueling eleven-round contest of the world's best divers, and well on his way into the finals the next day. He had a commanding lead in the preliminary in his bid to win his third Olympic Gold Medal under immense pressure. A top-twelve finish would assure each diver a spot in the finals and those twelve divers would compete the very next day, starting all over again, performing the same eleven dives, with the top three finishers earning medals. Just another day at the office. Or so we thought. During the ninth round, Greg Louganis, the greatest diver the world had ever seen, was preparing to do a reverse two-and-a-half somersaults in the pike position. It is a fairly difficult dive, carrying a degree of difficulty of 3.0, but it's a dive he's performed thousands of times, many times for 10s. The dive was a mainstay in his normal repertoire of dives for years, and was a common dive that most all of the world's divers would perform that day.

Greg's approach and hurdle to the end of the springboard was one of the most elegant movements in all of sports. It was a mixture of grace, beauty, balance, and precision, but also contained the strength and sheer power that would literally elevate him to heights that few divers could match. In springboard competitions throughout the world for over a decade, he out-jumped every competitor, not by inches, but by feet. He was a great competitor and never held back. He would jump to the moon and go for 10s each and every time. But on a few occasions over the years, both in training and in competition, he had a habit on the reverse takeoffs group of not pushing his hips through on takeoff and "standing dives up a little straight." In fact, my dad had noticed him a bit straight up on the basic 301B, a reverse dive pike, earlier in training that day. The result would be a trajectory that would leave him too close to the board with his upper extremities. It would happen from time to time on the reverse takeoffs, not only to him, but to most divers. The result would be a dive that fails to move away from the springboard into proper distance and would be a little too close for comfort. If it happened in training, a diver might "bail out" or "hold on," sensing that they were too close to the board, but in an Olympic preliminary and with qualifications on the line, you do the best you can to manage the situation. This was one of those days.

But in the ninth round on that memorable day in August 1988, that approach was the beginning of a dive that would be the most memorable, viewed, replayed, and talked-about dive in Olympic diving history. Greg would land on the end of the board after completing his approach and hurdle, with the board fully depressed for maximum height, but leaning a few degrees back in his balance. It was one of those times that he would normally make a slight adjustment and an extra effort on takeoff to move his hips through the board and into safe distance. But he didn't. He jumped to roof but didn't make that small correction.

The result was a reverse two-and-a-half somersaults in the pike position that was "too straight up" and dangerously close to the board. As we all know, and are still reminded today from time to time today, Greg would hit the top of his head on the springboard. He would land in the water headfirst but in an awkward, crouched position. The crowd sat silent in the Jamsil Swimming and Diving Stadium in shock at what they had just seen. Greg would climb out of the pool quickly, and on his own power, and rush with my dad back into the training room area to be looked at by a doctor. He asked how long he had before his next dive, which ironically enough, was another reverse takeoff dive. He had twenty minutes to be diagnosed, cleaned up, and continue—or elect to withdraw.

The rest is history. But what many people didn't know, or don't recall, was that the two dives that remained for him after hitting his head were both reverse takeoffs. In fact, four of the eleven dives he performed were from the reverse group. In round ten, Greg would return to perform a 5337D, the internationally recognized dive description, for a reverse one-and-a-half somersaults with three-and-a-half twists. It was the best and highest-scoring dive of the preliminaries. Again in the eleventh round, Greg would perform another reverse takeoff dive, but this time it was the hardest dive in his list; a reverse three-and-a-half somersaults in the tuck position. He did it well and finished third in a difficult and mentally draining afternoon. To go back up on that board, twenty minutes later, and have to do two more reverse takeoffs, was incredible. In fact, it was heroic. Greg would sleep very little that night, tossing and turning about what lay ahead of him the next day.

That very next day, with the memory of that dive still in his memory, he would go on to win the gold medal in the men's three-meter final by 26 points, performing the same dives all over again in a performance that hopefully will be remembered forever. He would go on to win a gold in the men's ten-meter platform final later on in the week, winning by 1.14 points. It completed his double-double, winning back-to-back golds in two consecutive Olympic Games. It was one of the single greatest athletic achievements I had ever witnessed.

What was so shocking is that it was the world's greatest diver making a mistake for all to see. It happened on the world stage in front of millions of viewers. In reality, it was an epic display of courage by a great warrior and champion. It made him human in the eyes of the world, if only for a moment, and, at least in my mind, inspired others to not let adversity deter them; to know that you can get back up. Greg did. In twenty minutes.

This is an example of a diving accident. In reality, he hit his head on a moving object, walked away on his own, did another dive twenty minutes later, was not seriously injured, never had to go to the hospital, had dinner that night, returned to win an Olympic gold medal less than twenty-four hours later and another gold medal later that week in his final athletic competition. It happened in sixteen feet of water by a trained athlete. *It was a "diving accident" in the sport of diving.* He got four stitches.

May 30, 2010, was just another balmy South Florida day. Ruben Camargo, 41, was enjoying a gathering with his family, friends, and acquaintances at his West Fort Lauderdale condo. Ruben's wife and son were in attendance, as well as guests who had gathered at the condo for food, drinks, games, and fun. Everyone was having a good time.

In fact, the party had turned into an all-day affair. Guests were dispersed around the development with some at Ruben's condo while others had decided to enter the pool after the posted hours to lounge around and continue the party. Ruben had admittedly consumed four four-ounce tumblers of vodka and orange juice; a good amount even for his five-foot-eight, 185-pound frame. He was later to have tested to have a 0.12 blood alcohol limit; well over Florida's 0.08 legal limit to operate a motor vehicle.

As a resident of the condo since 2005, it was estimated that Ruben had been in the pool one to two times a month over that five-year time period and had been swimming in pool about seventy to eighty times. Earlier that day, according to his own deposition, he dove into the six-foot-deep area where "no diving" signs were posted on the deck. He often had warned his son not to dive into the pool at all. He knew the pool had a shallow area. When asked why he had warned his son not to dive into the pool, below is the exchange between the defense attorney for the condominium and Mr. Camargo:

Attorney: Why is it dangerous to dive into the three-foot area of the pool?
Camargo: Because it is three feet.
Attorney: It's shallow?
Camargo: Yes.
Attorney: Someone can get hurt diving into three feet of water?
Camargo: Yes.

The following is the 48 seconds that result in an unnecessary and heartbreaking tragedy.

At 8:40:32 p.m., Mr. Camargo is shown on the condo's closed-circuit television surveillance camera with two others a few feet inside the gate after normal operating hours (see Figure 6.1). He is wearing a black sleeveless shirt, shorts, flip flops, and is holding a large white cup in his left hand.

At 8:40:36, Mr. Camargo then proceeds toward the pool (see Figure 6.2). The man standing at the corner of the pool in front of him is one of two men who will initiate the horseplay that will lead to Mr. Camargo diving into three feet of water.

At 8:40:48, Mr. Camargo is shown facing the camera with the white cup in his left hand (see Figure 6.3). In his deposition, he stated that he had no intention of swimming again that evening. But the two men shown behind him who initiated the accident had other ideas. Mr. Camargo barely had time to put his drink down on a table before being grabbed by the two men who will pull him toward the corner of the pool, ultimately leading to catastrophic injury for Mr. Camargo.

FIGURE 6.1 Mr. Camargo, holding the large white cup, stands with two others five feet inside the pool gate at 8:40:32 p.m.

FIGURE 6.2 Mr. Camargo walks toward the pool at 8:40:36.

FIGURE 6.3 Mr. Camargo is seen on the left edge of the frame at 8:40:48, preparing to put his white cup down.

FIGURE 6.4 At 8:41:16, Mr. Camargo is grabbed by two guests and pulled into the pool.

At the left side of the frame below, at 8:41:16, in the last frame prior to being pulled into the pool, we get a good view of the second man who initiated the accident (see Figure 6.4). He is in white shorts with his back to the camera. To his right is the second man. Mr. Camargo is barely visible between them in the scuffle that led to him being pulled to the edge of the pool, off balance, contorted on one leg, his arms entangled in theirs, then leaping into the three-foot, shallow end of the pool, headfirst, untrained, intoxicated with no plan or skills on how to protect himself. It was a recipe for disaster, brewing all day long, culminating in the 0.3–0.4 seconds it took for Mr. Camargo's head to strike the bottom of the pool, most likely in a "ducked" position, and change his life, and the life of his family, forever. The saddest thing is that it was preventable.

At 8:40:32, Mr. Camargo is shown walking into the pool area, at 8:41:20, 48 seconds later, Mr. Camargo is a quadriplegic who will never walk again. It happened in three feet of water by an untrained person.

Mr. Camargo's devastating loss of his ability to walk is irresponsibly and unfairly referred to in court and legal documents as a "diving accident." Moreover, the insurance industry classifies it as a diving accident, which has far-reaching effects on our sport and continues the myth that the sport of springboard and platform diving is dangerous. In fact, *Diving Injuries* (Gabrielsen et al., 2000) states that "competitive divers sustain very few injuries. To date, there is no record of any competitive diver receiving a neck fracture by striking the bottom of a pool that met the specifications promulgated by the various ruling bodies governing competitive diving. As the data in this book reveal, the recreational diver is at greatest risk."

The sport of springboard and platform diving is very safe. Diving into shallow water, by anyone, is not. It's not the sport of diving that creates unnecessary and excessive risk. The risk lies with untrained, unsupervised, intoxicated, negligent horseplay and bad judgments made by people entering shallow water or unsafe, unknown bodies of water while leading with their heads; with those people or entities associated with the design, planning, implementation, or operation of a facility where these injuries occur; and with the regulatory agencies and individuals charged with the responsibility to govern those facilities.

The point of this comparison is to help parents and others to understand that it is not the highly skilled or trained athlete that is sustaining paralysis or life-threatening injuries. The untrained recreational swimmer is at the greatest risk and shallow water is the overwhelming, consistent threat in the majority of these accidents.

SO WHAT HAPPENED?

Mr. Camargo hit the bottom of that three-foot pool in less than half a second. What we did not show in the preceding pictures is the actual dive itself, in which a trained diver could have escaped unscathed. But the untrained diver, under the influence of alcohol, in a split second of distress, will do just what Mr. Camargo did in the preceding frames after he hovered just a few feet above the water. He ducked his head, altering his body position from a "flat" horizontal, low-angle dive, into a high-angle, pike position trajectory that put him at 45-degree angle and sent him on a path directly into the bottom of the pool. There is a point in the video replay which I viewed over and over again where Mr. Camargo is suspended in air in that horizontal position. I would stop the recording each time, knowing that was the exact moment where he made that split-second, ill-fated decision that changed his life. You see, when diving into shallow water (water that is five feet or less), *the head position is of vital concern.* Inexperienced or "self-taught" swimmers will drop their head upon entry to absorb the impact on the top of their head, initiating a downward rotation and a deep water trajectory into very shallow water. The results of that are catastrophic.

His body is shown a few frames later entering the three-foot area of the pool in a high-angle dive and ending with an abrupt stop when his head hits the bottom, leaving him in a suspended handstand; his lower extremities sticking out of the water in a grotesque and disturbing ending. The compressive forces of a 180-pound object like Mr. Camargo impacting that pool bottom at a speed ranging from 0 to 12 feet per second at a 45 degree angle would range from 0 to 4700 pounds. A force of 100 to 1000 is enough to cause a quadriplegic injury. A force of 1700 pounds will fracture your skull. My guess is that when you just walked seconds before, then suddenly can no longer feel your legs and are lying in a heap in three feet of water, the science doesn't matter much. Life for Ruben Camargo, his young son, and wife had changed in an instant. That final frame is still engrained in my mind, and I am left with a vivid memory that I still think about. But Mr. Camargo is left paralyzed.

WHAT IS A SPINAL CORD INJURY (SCI)?

Spinal cord injuries (SCIs) are caused by a trauma to the vertical column. This trauma effects the spinal cord's ability to send and receive messages from the brain to the body's systems that control sensory, motor, and autonomic functions below the level of the injury itself. The autonomic nervous system is a control system that acts largely unconsciously and regulates bodily functions such as the heart rate, digestion, respiratory rate, pupillary response, urination, and sexual arousal (see Figure 6.5).

The spinal cord itself is about eighteen inches long. It extends from the base of the brain to down near the waist. Many of the bundles of nerve fibers that make up the spinal cord itself contain upper motor neurons (UMNs). Spinal nerves that branch off

Headfirst Recreational Aquatic Accidents

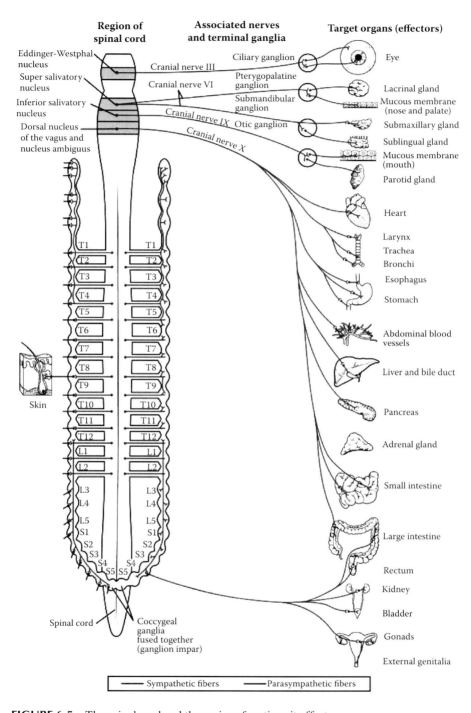

FIGURE 6.5 The spinal cord and the various functions it affects.

the spinal cord at regular intervals in the neck and back contain lower motor neurons (LMNs). The spine itself is divided into four sections, not including the tailbone:

- Cervical vertebrae (1–7), located in the neck
- Thoracic vertebrae (1–12), in the upper back (attached to the ribcage)
- Lumbar vertebrae (1–5), in the lower back
- Sacral vertebrae (1–5), in the pelvis

The severity of a spinal cord injury depends on the part of the spinal cord that is affected. The higher the SCI on the vertebral column, or the closer it is to the brain, the more effect it has on how the body moves and what one can feel. More movement, feeling, and voluntary control are generally present with injuries at lower levels. The two main types of SCIs are

- Tetraplegia (also known as quadriplegia), which is a direct result of injuries to the spinal cord in the cervical, or neck region, with loss of muscle strength associated with all four extremities.
- Paraplegia is a direct result of injuries to the spinal cord in the thoracic or lumbar areas which results in paralysis of the legs and lower parts of the body.

Without delving too deep into the science, one can have a *complete spinal cord injury* that produces total loss of all motor and sensory function below the level of injury. Nearly 50 percent of all SCIs are complete. Both sides of the body are equally affected. Even with a complete SCI, the spinal cord is rarely cut or transected. More commonly, loss of function is caused by a contusion or bruise to the spinal cord or by compromise of blood flow to the injured part of the spinal cord. A person can also sustain an incomplete SCI where some function remains below the primary level of the injury. A person with an incomplete injury may be able to move one arm or leg more than the other, or may have more functioning on one side of the body than the other.

Mr. Camargo suffered an injury to the C-5 and C-6 vertebrae. He is confined to a wheelchair for the remainder of his life, or until a cure is found. He and his family have incurred medical bills into the millions and their lives have been forever altered in a matter of seconds. As an expert in this case, it was my belief that Mr. Camargo was negligent for his role in the accident. He was intoxicated. He knew the pool was shallow and all parties entered the area illegally, fully aware that the pool was closed. However, the two men who pulled him Mr. Camargo into the pool that night were also highly irresponsible, negligent, and at fault. They walked away unharmed that night while Mr. Camargo laid in a local hospital in total and unimaginable fear. Had those two men not initiated the horseplay, usually reserved for children, Mr. Camargo would still be walking today. The two men were not named in any lawsuits as they had no assets.

The Statistics

According to multiple sources, the estimated statistics for all types of spinal cord injuries (SCIs) in the U.S. since 2010 include the following:

Headfirst Recreational Aquatic Accidents 111

- There are an estimated 17,000 new spinal cord injuries in the United States each year.
- There are approximately 276,000 people living in the U.S. today with SCIs. This number could be as many as 347,000.
- Since 2005, the average age at injury is 42 years old (up from 29 decades ago).
- Males make up 80 percent of new spinal cord injury cases.
- 52 percent are single at the time of the accident.
- The race/ethnicity factors:

White	63.5 percent
Black	22 percent
Hispanic	11 percent
Asian	2 percent
Other	1 percent
Native American	0.5 percent

- Since 2005, the category of injury upon hospital discharge is

Incomplete quadriplegia (tetraplegia)	45 percent
Incomplete paraplegia	21.3 percent
Complete quadriplegia (tetraplegia)	13.3 percent
Complete paraplegia	20 percent

- The causes of SCIs are

Auto accidents	38 percent
Falls	30.5 percent
Violence/gunshots	13.5 percent
Sports	9 percent (66 percent are headfirst recreational aquatic accidents)
Medical/surgical	5 percent
Other	4 percent

- There are an estimated 850 spinal cord injuries a year as a result of headfirst aquatic diving accidents.
- A 1998 study revealed that alcohol was involved in 49 percent of 196 people interviewed who had been injured in a swimming pool incident.
- The age of the U.S. population has risen, but the highest per-capita rate of injury occurs between ages 16 and 30. Young people are at the greatest risk.
- Of the 9 percent of sports-related SCIs, 66 percent, or two-thirds, are related to recreational headfirst water activities by untrained participants or in an unsafe environment (improperly labeled as diving).
- In a study of 440 SCIs in pool-related, recreational, headfirst water-related accidents, 93 percent occurred in water less than five feet deep.
- The National Swimming Pool Foundation reports that nine out of ten headfirst aquatic swimming accidents also labeled as "diving" injuries occur in water that is less than six feet deep.

When we consider the statistics and how they relate to aquatic-related, head-first diving accidents, it is clear that *shallow water* poses the greatest threat and the untrained, recreational diver is at greatest risk to be injured by their own or some other party's or entity's negligence.

The Facts

As mentioned earlier, one of the best studies by one of the most widely known authorities on diving accidents is found in the book *Diving Injuries* (Gabrielsen et al., 2000). The book will be cited numerous times moving forward, and much of its data analyzed, as it focuses strictly on recreational diving accidents from 1970–1989: who they happen to, what the causes are, when they happen, where they happen, why they happen, and how they happen as well as the many other factors involved in these accidents. His work and research, and its detail, is a benchmark for the knowledge, biomechanics, and general knowledge of recreational diving–related accidents. There are numerous places where recreational aquatic diving accidents occur. Again, study after study has revealed that the vast majority happen in *shallow water*.

First, we will look at the data from Gabrielsen's study of 601 accidents, which was broken down into the following categories:

1. Aboveground pool accidents—105 accidents
2. Inground pool accidents—211 accidents
3. Springboard/jumpboard accidents—92 accidents
4. Starting block accidents—32 accidents
5. Natural environment accidents—161 accidents (lakes, rivers, ponds, streams, quarries, oceans, embayments, reservoirs, ponds, and other water areas)

Data was used from accident cases that were in or had been in litigation where an expert witness was involved. Gabrielsen found early in his studies that this was the best, most factual, most reliable way to gather information on these accidents. Of the 601 cases, each of which became an independent study, 440 were classified in the first four categories and 161 were classified in the "natural environment" category.

We will look at data from each category, what can be learned from each, and finally, patterns that we see in recreational diving accidents. Rather than analyzing the individual cases themselves, a review of the overall data offers us a pattern that is quite telling. We will start with the entire data collected from the 440 accidents that occurred in manmade aquatic environments (inground, aboveground, springboards/jumpboards, and starting blocks). It is worth noting before we delve further into the numbers that of the 440 cases studied into pools, 93 percent (410) of them occurred in water less than five feet deep.

Below are some findings and patterns from the data collected on the 440 head-first recreational aquatic "diving" accidents that were studied in four categories (105 aboveground pool accidents, 211 inground pool accidents, 92 springboard/jumpboard accidents, 32 starting block accidents).

Headfirst Recreational Aquatic Accidents

Of the 440 headfirst aquatic accidents studied by Gabrielsen and his editorial board:

- 81.6 percent of the victims were male (359 males and 81 female).
- 89 percent of the injured males were between 13 and 33 years of age.
- 88 percent of the injured females were between 13 and 33 years of age.
- 53 percent of the male victims were five-foot-ten to six feet.
- 54 percent of the female victims were five-foot-four to five-foot-six.
- 45 percent of the male victims weighed 170–189 pounds.
- 75 percent of the female victims weighed 100–139 pounds.
- 60 percent of the male victims had C-5 to C-6 cervical damage.
- 62 percent of the female victims had C-5 to C-6 cervical damage.
- 63 percent of the entire group learned to swim between the ages of five and eight years old.
- 81 percent of the entire group cited their source of diving instruction as "self-taught" or from a "parent or family member."
- 51 percent of the entire group's accidents occurred in a "residential" pool.
- 52 percent of the entire group's accidents occurred in pool with no depth markings.
- 81 percent of the entire group's accidents occurred in pools with no bottom markings.
- 88 percent of the entire group's accidents occurred in pools with no lifeline in place.
- 46 percent of the entire group's accidents occurred in depths between three-foot-one and three-foot-six.
- In 80 percent of the entire group's accidents the water was clear.
- In 97 percent of the entire group's accidents the victim was a guest at the residence where the accident occurred.
- In 62 percent of the entire group's accidents it was the victim's first visit to that pool.
- In 61 percent of the accidents it was the victim's first dive into the pool that day.
- In 40 percent of the entire group's accidents some form of alcohol had been consumed within a twelve-hour period.
- 84 percent of the entire group's accidents occurred in pools where no warning signs were posted.
- 76 percent of the entire group's accidents occurred in pools with no lifeguard.
- In 75 percent of the entire group's accidents "friends" made the rescue.
- In 86 percent of the entire group's accidents no spineboard was used in the rescue.
- In 87 percent of the entire group's accidents no CPR was administered.
- 98 percent of the entire group's accidents occurred in water less than six feet deep.
- Of the 440 accidents, 20 percent (92) of the injured were diving from a springboard or jumpboard and *all* of them struck the upslope of the pool

bottom. This is the point where the deepest portion of the pool transitions to the shallower areas. These pool bottoms are classified as either "spoon-shaped" or "hopper-bottom" and in the author's opinion, are very dangerous pool designs, especially to the untrained.
- Accounting for all the various points of impact in all 440 cases, 76 percent (337) were at a depth of four feet or less.

The last and final statistic foreshadows what we are about to uncover; that time and time again, the majority of SCIs occur in shallow water. My definition of shallow water refers to water that is five feet or less. Remember, competitive diving takes place in pools which meet the necessary standards by governing bodies, but recreational diving accidents by the untrained participant and oftentimes not properly managed or supervised can occur in any number of environments. The recreational diver generally underestimates the dangers, hazards, and circumstances related to diving from a low height into shallow water. They are untrained and will overestimate their own agility, ability, and skills, as well as the looming risk. The recreational diver is at greatest risk. The primary objective always should be to avoid hitting the bottom. Therefore, the "default" recommendation is never to dive into water less than five feet deep. Also, a swimmer should never dive headfirst into any body of water of unknown depth, where the water is murky, or where the diver doesn't feel they have ample room to maneuver. It only takes a split second. Remember "feet first, first."

Now we will turn our attention to the individual categories themselves.

105 DIVING ACCIDENTS INVOLVING ABOVEGROUND POOLS

As of 1999, there were an estimated 6 million aboveground swimming pool in the United States. These types of pools became popular due to the ease and quickness of assembly as well as the low cost, easy maintenance, and the limited permit requirements. The companies that sell these pools rarely manufacture any of the parts themselves but rather put together attractive packages and options for the buyer and presto, a backyard pool.

The data on these types of pools indicates that they are inherently dangerous. The depths; the "invisible bottom," as seen in Figure 6.6, with no bottom markings; the design; and the surrounding decks all pose potential dangers to its users. While 95 percent of dives into these pool are by children, its victims are typically a bit older. The data from the 105 victims states that

- 80 percent of the victims were male.
- 20 percent of the victims were female.
- 79 percent of the injured males were between 13 and 30 years of age.
- Males between 19–21 years old were at greatest risk (25 victims).
- 90 percent of the injured females were between 13 and 30 years of age.
- 76 percent of the male victims were five-foot-ten or taller.
- 100 percent of the female victims were less than five-foot-eight tall.
- 59 percent of the male victims weighed 170 pounds or more.
- 85 percent of the female victims weighed 110–130 pounds.

Headfirst Recreational Aquatic Accidents

FIGURE 6.6 An aboveground pool, although inexpensive and convenient, can present some dangers to its users.

- 86 percent of the male victims had C-4, C-5, or C-6 cervical damage.
- 85 percent of the female victims had C-4, C-5, or C-6 cervical damage.
- 95 percent of the entire group learned to swim by eight years old.
- 85 percent of the entire group had no formal diving instruction and were "self-taught" or from a "parent or family member."
- 100 percent of the entire group's accidents occurred in an aboveground "residential" pool.
- 95 percent of the entire group's accidents occurred in pool with no depth markings.
- 100 percent of the entire group's accidents occurred in pools with no bottom markings.
- 100 percent of the entire group's accidents occurred in pools with no lifeline in place.
- 100 percent of the entire group's accidents occurred in aboveground pools with vinyl liners.
- In 98 percent of the entire group's accidents the victim was a guest at the residence where the accident occurred.
- In 66 percent of the entire group's accidents it was the victim's first visit to that pool.
- In 69 percent of the entire group's accidents it was the victim's first dive of the day.
- In 63 percent of the entire group's accidents it was caused by a dive from the attached deck.
- In 41 percent of the entire group's accidents some form of alcohol was involved.
- In 96 percent of the entire group's accidents no rules were posted.
- 100 percent of the entire group's accidents occurred in pools with no lifeguard.
- In 95 percent of the entire group's accidents "friends" made the rescue.

- In 94 percent of the entire group's accidents no spineboard was used in the rescue.
- Aboveground pools are inherently shallow. Diving into aboveground pools is dangerous and should not be done from a board, deck, running, from the ground or under any circumstances.

What can we learn from aboveground pool accidents?

1. The overwhelming majority of victims are guests at the residence.
2. The diving accidents can occur in numerous different ways. Of the 105 studied here, there were 12 different scenarios or locations around the pool itself that presented a danger.
3. Not one victim knew of the risk or that they could break their neck.
4. Warning signs or depth markers are rarely posted due to the temporary nature of most of these pools and the fact that they are at a residence with little or no supervision.
5. Most of the victims are first-time visitors to the residence.
6. The vinyl covering that acts as the pool bottom is slippery, so naturally a victim's hands will slip upon impact, exposing their head and neck, leaving them no protection from impact.
7. The majority of the time the home or pool owner was not present.
8. The majority of victims had no formal diving instruction.

211 DIVING ACCIDENTS INVOLVING INGROUND POOLS

Below are findings and patterns on data from the 211 victims of inground pool diving accidents that Gabrielsen and his editorial board studied. Again, shallow water and untrained participants are the culprits. These pools were located in family backyards, apartments, condos, hotels, and motels. See Figure 6.7.

Below are some findings and patterns from the data collected by Gabrielsen:

- 80 percent of the victims were male.
- 20 percent of the victims were female.
- 81 percent of the injured males were between 13 and 30 years of age.
- 71 percent of the victims were single.
- 80 percent of the injured females were between 13 and 30 years of age.
- 51 percent of the male victims were between five-foot-ten to six feet.
- 30 percent of the female victims were taller than five-foot-six.
- 45 percent of the male victims weighed 170–189 pounds.
- 73 percent of the female victims weighed 129 or less.
- 57 percent of the male victims had C-5 and/or C-6 cervical damage.
- 58 percent of the female victims had C-5 and/or C-6 cervical damage.
- 90 percent of the entire group learned to swim by eight years old.
- 79 percent of the entire group had no formal diving instruction and were "self-taught" or taught by a "parent or family member."

Headfirst Recreational Aquatic Accidents

FIGURE 6.7 A recreational swimmer entering an inground pool headfirst and unprotected.

- 33 percent of the entire group's accidents occurred in an inground "residential" pool.
- 26 percent of the entire group's accidents occurred in an inground "apartment/condominium" pool.
- 25 percent of the entire group's accidents occurred in an inground "motel/hotel" pool.
- There were 13 different pool designs where these 211 accidents occurred.
- 39 percent were gunite basin composite pools.
- 98 percent of the accidents were in water less than five feet deep.
- 63 percent of the pool had depth markings.
- 79 percent did not have bottom markings.
- 97 percent of the entire group's accident victim were a guest at the residence, apartment, condominium, or motel/hotel where the accident occurred.
- In 66 percent of the entire group's accidents it was the victim's first visit to that pool.
- In 65 percent of the entire group's accidents it was the victim's first dive of the day.
- In 71 percent of the entire group's accidents they dove into a shallow area of the pool.
- In 48 percent of the entire group's accidents some form of alcohol or drugs were involved.
- In 64 percent of the entire group's accidents rules were posted.
- In 91 percent of the entire group's accidents no warning signs were posted.
- 75 percent of the entire group's accidents occurred in pools with no lifeguard.
- In 73 percent of the entire group's accidents "friends" made the rescue.

- In 92 percent of the entire group's accidents no spineboard was used in the rescue.
- 87 percent of the accidents occurred from the victim hitting the bottom at four feet or less.
- Nearly nine-tenths of the accidents in inground pools occurred in water less than four feet deep.

What can we learn from aboveground pool accidents?

1. The overwhelming majority of accidents occur in less than five feet of water.
2. Apartment/condo and hotel/motel pools generally operate without lifeguards.
3. There are certain types of pool bottom contours that present more dangers than others.
4. Most accidents occur on the victim's first visit to the pool.
5. Most victims were unaware of the risk.
6. Residential pool owners must be actively present, provide instruction, and provide warnings to their guests about depth and bottom contour, and must enforce a "no diving" policy.
7. Warnings, rules, and depth markers must be posted at all types of pools.
8. Bottom markers are helpful to see the bottom and its contour.

92 DIVING ACCIDENTS INVOLVING A SPRINGBOARD/JUMPBOARD

Next, we will examine data collected as a result of 92 swimmers who got injured diving off springboards or jumpboards into inground swimming pools. For clarification, a springboard can be classified as the name implies, a springy board, and may or may not be a competitive springboard but is any board that gets its bounce from its own inherent flexibility (see Figure 6.8). A jumpboard, on the other hand, is a board

FIGURE 6.8 A sixteen-foot competition springboard.

Headfirst Recreational Aquatic Accidents

FIGURE 6.9 A backyard jumpboard.

that you would find in a small backyard or hotel pool and it has a "cantilever" or coil springs to make it springy (see Figure 6.9).

Without delving into the science, there are three things that will affect a dive from either of these types of boards:

1. The human factors.
2. The environmental factors.
3. The mechanics of the dive.

Below are some findings and patterns from the data collected by Gabrielsen and his editorial board of experts:

- 93 percent of the victims were male.
- 7 percent of the victims were female.
- 63 percent of the injured males were between 16 and 24 years of age.
- 83 percent of the injured females were between 10 and 15 years of age.
- 58 percent of the male victims were between five-foot-ten and six-foot-one.
- 100 percent of the female victims were under five-foot-five.
- 53 percent of the male victims weighed 160–189 pounds.
- 100 percent of the female victims weighed 139 pounds or less.
- 72 percent of the male victims had C-4,5, C-5, and C-5,6 cervical damage.
- 92 percent of the entire group learned to swim by eight years old.
- 82 percent of the entire group had no formal diving instruction and were "self-taught" or taught by a "parent or family member."
- 56 percent of the entire group's accidents occurred in an inground "residential" pool.
- 67 percent of the accidents occurred in a traditional, rectangular-shaped pool.

- 52 percent of the accidents occurred in "spoon" bottom pools.
- 48 percent of the accidents occurred in "hopper" bottom pools.
- 60 percent of the pool had no depth markings.
- 92 percent did not have bottom markings.
- 86 percent of the accidents occurred in clear water.
- In 93 percent of the entire group's accidents victims were a guest at the residence, apartment, condominium, or motel/hotel where the accident occurred.
- In 54 percent of the entire group's accidents it was the victim's first visit to that pool.
- In 54 percent of the entire group's accidents it was the victim's first dive of the day.
- In 42 percent of the entire group's accidents some form of alcohol or drugs were involved.
- In 43 percent of the entire group's accidents rules were posted.
- In 93 percent of the entire group's accidents no warning signs were posted.
- 78 percent of the entire group's accidents occurred in pools with no lifeguard.
- In 76 percent of the entire group's accidents "friends" made the rescue.
- In 93 percent of the entire group's accidents no spineboard was used in the rescue.
- In 89 percent of the entire group's accidents paramedics were involved.
- 70 percent of the cases involve an impact in water less than five feet deep.

It should be noted that *none* of the pools affiliated with these accidents met the specifications of

- United States Diving (the national governing body for competitive diving in the United States and recognized by the United States Olympic Committee)
- National Collegiate Athletic Association (the national governing body for collegiate athletics)
- Federation Internationale De Natacion (the governing body for world, international, and Olympic diving events recognized by the International Olympic Committee)
- National Federation of High School Athletics (the national governing body of high school athletics)

What can we learn from springboard/jumpboard accidents?

1. Bottom configuration in both the "hopper" and "spoon" type pools where depths and subsequent impact "zones" vary, only have one deep area which presents dangers to any recreational diver or swimmer.
2. It is my opinion that backyard, recreational diving boards of multiple varieties present more danger and risk than value.

3. Depth markers and diving rules are rarely posted adequately.
4. Pools are not deep enough in the majority of locations where recreational boards are mounted.
5. Adult participants present even more dangers because of height and weight factors.
6. The upslope of the pool bottom, oftentimes too close to the board, is dangerous for any type of headfirst diving activities from a board or even the side of the pool.
7. "Hopper" and "spoon" type pool designs are unsafe for diving board installation of any kind.
8. Males are at greatest risk.
9. Most victims are guests.

32 DIVING ACCIDENTS FROM SWIMMING STARTING BLOCKS

The 32 accidents we will look at data from are the result of someone diving from a starting block into an inground swimming pool. Both the starting block (see Figure 6.10) and the types of dives swimmers performed off the blocks to begin a race have evolved dating back to the 1920s. Various heights were permitted for the flat-style starting blocks ranging from a height of 18 to 30 inches. In 1971, the sloped starting block was introduced but at not more than a 10-degree angle. Swimmers adapted throughout the years, employing different techniques such as the "flat" dive and the "pike" dive and varying degrees of entry into the water were explored. The "pike" dive required much more precision and ability by the swimmer to level off, or plane, underwater at a depth of two-and-a-half to three feet or risk hitting the bottom due to the higher angle. Obviously, this type of maneuver required practice and proper

FIGURE 6.10 A starting block for swimming.

planning ahead of time. "Plan your dive" is always a mantra to be heeded in any situation.

Below are some patterns, findings, and comments from the data collected by Gabrielsen:

- 59 percent of the victims were male.
- 41 percent of the victims were female.
- 89 percent of the injured males were between 13 and 18 years of age.
- 76 percent of the injured females were between 13 and 18 years of age.
- 57 percent of the male victims were between five-foot-ten and six feet.
- 69 percent of the female victims were five-foot-four or less.
- 52 percent of the male victims weighed 170–189 pounds.
- 76 percent of the female victims weighed 129 pounds or less.
- 94 percent of the male victims had C-4, C-4,5, C-5, and/or C-5,6 cervical damage.
- 76 percent of the female victims had C-4, C-4,5, C-5, and/or C-5,6 cervical damage.
- 87 percent of the entire group learned to swim by eight years old.
- 75 percent of the entire group had no formal diving instruction and were "self-taught" or taught by a "parent or family member."
- 68 percent of the entire group's accidents occurred in a high school or college pool.
- 84 percent of the entire group's accidents occurred in a rectangular pool.
- 100 percent of the entire group's accidents occurred in a concrete pool.
- 100 percent of the pools had depth markings.
- 100 percent of the pools did not have bottom markings.
- 93 percent of the entire group's accidents occurred in pool with clear water.
- 53 percent of the entire group's accidents occurred during a practice session.
- 59 percent of the entire group's accidents occurred while doing a "pike" dive. Only 12 of the 19 who were injured had received any instruction from the coach.
- 84 percent of the accidents were by members of the junior high or high school swim team.
- In 37 percent of the entire group's accidents it was the victim's first visit to that pool.
- In 100 percent of the entire group's accidents no form of alcohol or drugs were involved.
- In 100 percent of the entire group's accidents rules were posted.
- In 100 percent of the entire group's accidents no warning signs were posted.
- 100 percent of the entire group's accidents occurred in pools with no lifeguard.
- In 100 percent of the entire group's accidents no spineboard was used in the rescue.
- 100 percent of the 32 cases reported took place in water less than four feet deep.

Headfirst Recreational Aquatic Accidents

What can we learn from swimming starting block diving accidents?

1. A water depth under the blocks of three-and-a-half feet was a major cause of accidents.
2. The majority of victims were high school-aged or younger.
3. Victims did not receive adequate training especially when diving into extremely shallow water.
4. An untrained swimmer, executing a racing dive off of a 30-inch starting block into three-and-a-half to four feet of water in dangerous.
5. The "pike" racing dive is of particular risk to an untrained swimmer in shallow water.
6. The water under starting blocks should be at least five feet deep.

161 DIVING ACCIDENTS IN THE NATURAL ENVIRONMENT

Half of all diving injuries come as a result of people diving into lakes, rivers, ponds, streams, quarries, oceans, embayments, reservoirs, ponds, and other water areas. Common sense would suggest that these contain murky water where seeing below the surface or being able to discern depth is impossible (see Figure 6.11). One should never dive into any body of water when going into it for the first time and certainly not if the depth is unknown or objects below the surface are unknown. These bodies of water are under the jurisdiction of many different entities, from the federal government to state, county, city, and private organizations. Not surprisingly, most if not all of these accidents could have been prevented by better education, understanding, and appreciation of the dangers of diving headfirst into any body of water. Individuals must take responsibility, but the entities mentioned above must also act to properly warn, educate, eliminate, prevent, and supervise these areas.

FIGURE 6.11 Diving from piers into murky ocean or lake water can be catastrophic.

My personal opinions and observations from the data are as follows:

- 91 percent of the victims were male.
- 9 percent of the victims were female.
- 81 percent of the injured males were between 13 and 24 years of age.
- 76 percent of the injured females were between 13 and 24 years of age.
- 85 percent of the male victims were between five-foot-eight to six feet.
- 100 percent of the female victims were five-foot-seven or less.
- 33 percent of the male victims, the majority, weighed 170–179 pounds.
- 76 percent of the female victims weighed 110–129.
- 72 percent of the male victims had C-5 or C-5,6 cervical damage.
- 46 percent of the female victims had C-5 or C-5,6 cervical damage.
- 32 percent of the entire group's accidents occurred in a property owner by a community/corporation.
- 55 percent of the entire group's accidents occurred in a lake, while 19 percent of the entire group's accidents occurred in a river. 74 percent of the accidents occurred in a lake or river.
- Florida and California accounted for 37 percent of the cases by themselves.
- 45 percent of the entire group's accidents occurred in designated swim areas.
- 55 percent of the entire group's accidents occurred in non-designated swim areas.
- In 81 percent of the entire group's accidents the water was turbid or cloudy.
- 80 percent of the entire group had no formal diving instruction and were "self-taught" or taught by a "parent or family member."
- In 67 percent of the entire group's accidents it was the victim's first visit to the site.
- In 80 percent of the entire group's accidents it was their first dive that day.
- There were 23 various activities leading to these accidents.
- 55 percent of the accidents occurred in lakes.
- Florida had the most accidents and more than double the next state.
- In 46 percent of the entire group's accidents no form of alcohol or drugs were involved.
- In 65 percent of the entire group's accidents no rules were posted.
- In 92 percent of the entire group's accidents no warning signs were posted.
- 81 percent of the entire group's accidents occurred in pools with no lifeguard.
- In 93 percent of the entire group's accidents no spineboard was used in the rescue.
- 16 percent of the cases occurred in water less than two feet deep.
- 85 percent of the entire groups accidents occurred in water four feet deep or less.

What can we learn from diving accidents in the natural environment?

1. The young, overly aggressive male recreational diver is at greatest risk because he overestimates his ability and misjudges the depth and conditions.
2. The victims are usually self-taught novices with no formal instruction whatsoever.

3. Most accidents occurred on the very first visit to the location.
4. Most accidents were in water less than four feet deep.
5. Warning, depths, and rules were rarely posted to adequately discourage the public.
6. Piers and docks are an attraction for horseplay and a severe hazard to your health.
7. Cloudy or turbid water hides objects underwater as well as the depth.
8. Property owners who have water on, in, or around their property routinely fail to warn of impending dangers and prohibit diving in certain areas.
9. Property owners fail to supervise or provide instruction where swimming is permitted.
10. Owners and operators of facilities fail to properly inform and warn the public.
11. Lifeguards were rarely on duty in designated swimming area.
12. Property owners allow swimming in high-risk areas and fail to perform proper risk analysis.
13. Owners fail to remove debris, trees, branches, or potentially dangerous equipment.
14. Property owners build or misplace structures over or near water, creating a dangerous environment.
15. Running dives into the ocean or breaking waves going headfirst into the water is extremely dangerous.
16. The general public must be better educated on the inherent dangers of diving, shallow water, and learn to enter water "feet first, first."

The biggest diving dangers in the natural environment:

1. Cliffs or embankments near rivers and lakes
2. Bridges
3. Ocean piers
4. Boat docks on lakes or rivers
5. Retaining walls or bulkheads
6. Boats in shallow water
7. Rope swings

THE NUMBER-ONE CAUSE OF HEADFIRST RECREATIONAL AQUATIC ACCIDENTS: SHALLOW WATER

Shallow water is the enemy. That is not to put all responsibility on the participants or the victim, or to absolve all the other parties that may be liable in the event of an accident. But the general public's level of education, and having a general knowledge and understanding that shallow water is the enemy, would go a long way toward preventing more catastrophes.

It is a common theme throughout this chapter that time and time again, no matter the situation, environment, apparatus, setting, age, sex, height, weight, pool design, location, or any of the many pieces of data collected: Diving into shallow water will put you in a wheelchair, or worse. Of course, it is the responsibility of the pool owners, property owners, agencies, and other affiliated parties to post the necessary warnings, rules, and dangers as well as to properly educate the public that a danger does exist.

The subject of diving safety is the topic of focus, discussion, and teaching at the first class of every Learn to Dive class I have ever taught and the very essence of USA Diving's teaching program at the beginner level. It is the focus on day one and class number one. It is of immense importance that children know to enter the water cautiously for the first time with their feet, not their head, and to be respectful of the dangers that exist. We teach the mantra "feet first, first" when entering a new pool for the first time; knowing the depths of the pool, the posted and unposted rules, the markings, and familiarizing oneself with the immediate environment and potential dangers that can exist in any pool. This caution and attention to safety, detail, personal responsibility, respect for the water, respect for equipment, and other key lessons on protecting oneself in that environment are essential to safe diving moving forward. This approach gives children in the Dive Safe, Learn to Dive, and Flip-n-Fun Diving Lessons that we have taught a healthy and necessary respect for the water. Our lessons were given in a world-class facility, with eighteen feet of deep water everywhere while keeping in mind that our students would return to residential pool of their own, or their friends', of much less depth. This same approach must certainly be adhered to in other facilities where the water is much shallower. Remember, "feet first, first."

Divers in beginning lesson classes learn the very basics and safety of a dive from the side of the pool. Our lessons were taught in eighteen feet of water! They learn what depths to avoid and are taught not to dive into shallow areas of the backyard pools that are so prevalent in Florida. They learn the proper position of the head and arms overhead for safe diving as well as proper body alignment. Because novice divers will tend to duck their head upon impact, a certified and knowledgeable USA Diving coach will teach beginners in organized lessons to *plan their dive*, how to keep their eyes on the surface of the water as it approaches, how to grab their hands overhead covering their ears with their shoulders for entry, and how to "scoop" out and away into safety. As competitive divers progress throughout a lesson plan, they will become adept at using more of the water's depth and going deeper when in a competitive diving pool, but the initial teachings to beginners are "survival" techniques in shallow water. Most importantly, they are ingrained with the idea that they be mindful of the depth of the pool, plan their dive, and *always allow for an adequate and safe amount of water in which to maneuver.*

On the contrary, an untrained diver at any age is not aware of the many dangers that exist in the water. A child who has participated in a USA Diving lesson program will learn lifetime skills that may save them from serious injury and become second nature forever. I urge all parents to have their children take beginning diving lessons at a USA Diving certified program so they can learn the basics of safe diving and avoid or manage the pitfalls that they could be exposed to at an impromptu gathering at a friend's backyard pool; a hotel pool; a lake, bridge, or pier in the natural environment; or any place that danger may exist. As parents, we can't always be there.

So once your child has learned to swim, you can get them into a Learn to Dive class when they are as young as four years old. Most kids won't become Olympic champions, but they can be part of a great, exciting, safe sport; gain a basic understanding; a healthy respect for the water and diving boards; and an ability to protect themselves in any water depth or aquatic environment.

Once again, for an untrained diver who unknowingly or freely dives into shallow water (water less than five feet), the head position is of particular concern. Inexperienced and untrained divers will drop their head upon entry to absorb impact on the top of their head causing a downward, rolling rotation, and in turn causing a deep water trajectory with potentially catastrophic results. The biggest factor associated with this is time. In shallow water, a person diving from even a low height at a 45-degree angle will hit the bottom of a three-and-a-half foot pool in less than half a second from the instant their head pierces the surface. In this scenario, if the hands are not grabbed properly, or at least overhead, and the arms are not locked and in a correct position to protect, deflect, or absorb the impact of hitting the bottom, the diver may be seriously injured. This is why the mantra of "plan your dive" is so vital. Because the time interval is so short, proper "programming" of the diver's trajectory must take place in advance. In other words, the diver must program the lifting of the head and the arching of the back immediately after entry into water too shallow or there may be significant consequences. As the old saying goes, "It's not the fall but the sudden stop," is true. When a diver's head hits the pool's bottom, the momentum is temporarily stopped, but the torso, containing the center of mass, will attempt to continue its motion. The bones in the neck are then "compressively loaded and the resultant force can be sufficient to cause irreversible injuries." A good visualization of this process is the result when a freight train engine suddenly stops and the cars try to continue their motion through the stopped engine. As we so often refer to from USA Diving's Dive Safe Manual:

> The analyses leave one with an inescapable conclusion. Divers must be given an adequate volume of water in which to finish their dives and maneuver to return safely to the water's surface. If an adequate water volume for safe diving cannot be provided, then the body of water may be suitable for swimming only. If so, then adequate warnings must be posted prohibiting diving.

HOW CAN WE REDUCE DIVING ACCIDENTS?

As in most situations when injuries occur, there are many fingers being pointed. But the reality is that all entities share the burden to provide a safe aquatic environment but to also educate, warn, and protect our children and anyone who enters a pool or a body of water in the natural environment. Clearly, we cannot monitor every lake, river, and stream, but helping people understand that shallow water and other aspects of the aquatic world are a risk to them is not only our responsibility but our duty. The first step in reducing the number of injuries in the aquatic environment by swimmers who decide to dive, leading to paralysis, is communicating to the public that there is risk involved and they could get seriously hurt. *The education starts when children are young so they have an understanding about the dangers present in the aquatic setting. It is greatly important to enroll children in swimming and diving lessons.*

In legal cases, the pendulum shifts from case to case. In some legal cases the failure of a pool owner, property owner, or agency to properly monitor, protect, and warn the public is obvious; other times the responsibility lies on the injured party for their part in the accident. But we all share the responsibility to provide a safe environment for those who want to dive into our pools and enjoy all that the water has to offer. It is the author's belief that just as all children should learn to swim as infants, they should also have access to, and take, diving lessons from a USA Diving–certified coach or instructor. Water is fun. It is also dangerous and can kill or seriously injure you whether you are swimming or diving. Learn to protect yourself in both environments. Individuals must do their own part in protecting their well-being. That being said, all the entities who participate in the various aspects aquatics must do their parts, including the following:

POOL DESIGN AND RECOMMENDED CHANGES FOR SAFER POOLS

When I was fourteen, my dad installed a beautiful pool in the backyard of our Mission Viejo, California, home complete with a competition one-meter springboard and separate Jacuzzi. The pool bottom had various contours and that one-meter was a model B Duraflex springboard that would rocket you into the air. It was a dream come true for us. My friends were all competitive divers from our team so everyone who came over were excellent swimmers and highly skilled divers who knew that there were risks, how to maneuver in the pool, how to safely use the springboard, and who were fully aware of the pool depths and depth changes. We never had one accident or anything that even resembled one. Trained swimmers and divers have a healthy respect for the water. Untrained swimmers and divers are unaware of the risks inherent in the aquatic world, and as a result, tend to take risks as they overestimate their ability and underestimate the looming danger. When untrained swimmers are using pools, we need to look at what designs are unsafe for diving boards.

Spoon, Hopper, and Vinyl Bottom Pools

When you combine an untrained swimmer with a springboard placed in a "spoon bottom" or "hopper bottom" pool, or a swimmer diving into an aboveground pool with a vinyl bottom, you have a recipe for disaster. Even if legislation were passed today to outlaw this specific type of pool design, there are still thousands throughout this country that exist in various locations. They present a danger to everyone in my opinion and in the opinion of those who study this topic. The sudden depth changes, the upslope of the bottom straight ahead of where swimmers dive from the springboard, and the winding contours of the pool bottom present dangerous consequences to uninformed swimmers.

In addition, the "vinyl bottom" design so prevalent in aboveground pools is unsafe, in my opinion, if participants insist on diving into those pools—as so many do, resulting in injury. Time after time, those injured stated that their hands slipped off the bottom, exposing their head and causing injury. In a pool where friction on the bottom is present, the arms extended overhead could possibly protect the head, avoiding catastrophic injury. Again, the swimmers should not be diving into an aboveground pool under any circumstances, but the vinyl pool flooring presents an added danger.

Generally speaking, more attention needs to be placed on pool designs by the industry as well as the operation, manufacturing, installment, supervision, and overall attention to safe diving into pools and how to better educate the public. Some general observations and opinions are below.

Markings on Pool Bottoms

It is the recommendation of many aquatic safety experts that markings on the bottom of the pool would help swimmers see, judge, and help visualize where the bottom of the pool is. In 81 percent of the 440 pool injury cases involving a swimmer entering the water headfirst, there were no such markings on the bottom of the pool (see Figure 6.12). In many scenarios, the trajectory of the diver is already determined, but being able to see the approximate location of the bottom of the pool by having large markers could help reduce injuries.

Starting Blocks

Starting blocks should not be placed over water less than six feet deep.

Water Slides

Water slides should not be placed in an area of the pool where the water is less than eight feet deep.

Depth Markers

Depth marker laws should be consistent in all states and among all pools, locations, and owners. The proper depth should be required to be visible on the edge/deck and on the interior walls of the pool above the water line.

FIGURE 6.12 A backyard pool with no depth markings, no bottom markings, and no posted rules.

Fluorescent, Solar or Neon Nighttime Depth Markers
Depth markers that can be easily seen at night when pools are open can help patrons be more aware of the depths and avoid further accidents.

Locations of Pool
Code should be passed at the state and local level so that pools are not placed in an area near a house, hotel, walkway, or structure that encourages jumping or diving into the pool (see Figure 6.13).

Warning Signs
Proper warning signs should be placed in and around all pools prohibiting diving in water less than five feet deep, the default "no diving" depth for the general public. Signs should warn users of potential dangers.

Lighting
Adequate lighting in and around pools that permit swimming at night should be enforced. Swimmers must be able to see the bottom of the pool.

Barriers
Self-closing and self-locking pool fencing around all pools should be required to prevent children from entering.

Basic Water Depth Awareness
Parents and swimmers of all ages should have an awareness of the very basic concept that water over people's heads is where people generally drown and water less than five feet deep is where the greatest risk of diving headfirst injuries occur and where people generally break their necks.

FIGURE 6.13 A beautiful setting but a temptation to dive from five meters into inadequate water depth.

Elimination of Safety Ledges
Safety ledges in the deep end should be discontinued as they present a hazard in deeper water.

Marked Edges or Seats
Wherever steps or edges exist on entering the pool, it is recommended that those areas be clearly marked.

Recessed Steps, Ladders, Anchors, and Receptacles
There should be no obstructions such as ladders or railings in the pool itself and all steps, ladders, anchors, and receptacles should be recessed in such a way that they aren't a danger to anyone.

Lifeline
All pools should be divided by a lifeline separating deeper water from water that is shallower. This visual reference is just another easy safety measure for swimmers to know that the water depth is changing.

Color-Coded Pool Edges
Require red and green paint along the coping of pools indicating "no diving" and "safe diving" areas.

Improved Instructional Signs for All Pool Types and Locations with Boards
In all pools where a springboard/jumpboard is present, require the following sign to be posted at that location.

Rules for Springboard or Jumpboards
Prior to using the board, you must

1. Feet first, first!
2. Know the water depth.
3. Know where the shallow areas are.
4. Visually check the shape of the pool bottom in and around the board.

Then follow these rules:

1. Know the depth directly under the board where you will land.
2. First time: Go off the board "feet first, first."
3. Plan your first dive knowing how you will maneuver safely.
4. Second time: Go off the board from a standing dive position with arms overhead and hands grabbed, squeezing your ears with your shoulders.
5. No double bouncing is allowed.
6. All jumps and dives must be done straight ahead, not to the side.
7. Only one person at a time on the board is permitted.
8. Check to see that the person who went before you is safely out of the way.
9. Always go off the board from *two* feet. No jumping or diving from *one* foot.

10. After entry, scoop out and up to the surface.
11. Once you're to the surface, swim to the ladder.
12. No rafts, tubes, or toys allowed on the board.
13. No horseplay, clown dives, or dangerous stunts allowed off the boards.
14. Do not try dives you have never done or had no instruction for.
15. Don't use the boards if you have had any alcohol or drugs of any kind.
16. Do not go off the board if you are significantly overweight.

Assuming the water depth is six feet or more and diving is allowed, the following sign is recommended reading for each swimmer. In addition, it is highly recommended that parents enroll their children as early as four years old in a USA Diving safety–certified Learn to Dive program once they have learned to swim in water that exceeds their own height. Realistically, not everyone will do so and children will continue to ignore the "no diving" signs and warnings around swimming pools. For that reason, the sign below may discourage those from diving or save those that do from serious injury:

Rules for Diving into the Pool from the Side
You should know that

1. Feet first, first.
2. Diving into water under five feet is unsafe.
3. Diving too deep can cause serious injury.
4. Swimmers have been paralyzed by hitting the bottom.
5. If you dive, you should plan to go barely under the water, then resurface.

Safe diving from the side requirements:

1. Know the water depth.
2. Know where the shallow areas are.
3. Visually check the shape of the pool bottom in and around the board.
4. Know the depth directly where you will land.
5. First time: Go "feet first, first."
6. Do standing dives only.
7. No running dives allowed.
8. Plan your first dive knowing how you will maneuver safely.
9. Pick a spot six feet out on the water and dive toward it in a flat position.
10. Once you enter the water, raise your head up, arch your back, push down with your hands, and resurface.
11. Do not dive if you have had alcohol or drugs of any kind.
12. Do not dive deep or you will hit the bottom. Low angle, 10–15 degree surface dives *only*.

BETTER POOL MANAGEMENT AND OPERATION

Better Supervision
Public and semi-public pools should have lifeguards on duty at all times.

Industry Standards

The National Spa and Pool Institute (NSPI) must address diving safety issues more fervently—especially as it pertains to shallow water, slides, springboard/jumpboards—and encourage parents nationwide to get their children into Learn to Dive programs provided by USA Diving.

State and Local Laws and Regulations

Stricter laws and more attention should be placed on residential pool owners.

OPERATION AND MANAGEMENT BY RESIDENTIAL POOL OWNERS

Of the 440 diving-related injuries in pools studied from the comprehensive book *Diving Injuries* (Gabrielsen et al., 2000), 83 percent happened in private residences, condominiums/apartments, and hotel/motel settings. As one can imagine, these environments have few lifeguards, posted rules and warnings, and little supervision. Gabrielsen and his editorial board came up with the following list of recommendations for the owner of residential pools:

1. Make sure your pool is free of debris or anything that can cause unnecessary injury.
2. Always supervise the use of your pool. You, and others, could be liable if someone is injured.
3. Make sure that people who use your pool have had swimming and diving instructions.
4. Give visitors to your home and pool a thorough safety talk on the pool and its rules. Go over the depth of the pool, board, spa, and slide rules.
5. Insist that first-time users enter the pool "feet first, first" or by the ladder and go underwater to scope out the bottom shape.
6. Have pool rules posted everywhere around the pool area.
7. Have pool depth markings posted around the coping of the pool.
8. As a precaution, have bottom markers so users can see where the pool bottom is located.
9. Make sure the pool water is clear.
10. Provide good lighting in and around the pool at night.
11. Make sure that adequate depth is provided wherever slides are placed.
12. Always have a first-aid kit and other safety equipment.
13. Take CPR and basic water safety courses.
14. Supervise, watch, and warn your guests if needed.
15. Prohibit diving in the shallow end or through inner tubes, and so on.
16. Hire a lifeguard whenever possible and especially for large gatherings of more than twenty-five.
17. Have safety gates and covers when not in use.
18. Protect yourself from liability by going above and beyond in regards to safety and protecting all of your guests.

Operation and Management by Hotel/Motel, Condo/Apartment, and Public Pools (in Addition to the Guidelines Above)

1. Have lifeguards on duty during open pool hours.
2. Secure the pool area with locked gates or covers when the pool is closed.
3. Have a closed circuit television monitoring system at all hours.
4. Have pool rules posted everywhere around the pool area.
5. Have pool depth markings posted around the coping of the pool.
6. As a precaution, have bottom markers so users can see where the pool bottom is located.
7. Make sure the pool water is clear.
8. Install good lighting if pool is open after dark.
9. Make sure pool operators know local and state requirements.
10. Have pool fencing a minimum of five feet in height and locked gates whenever the pool is closed.

Recommendations to the Insurance and Medical Industry

Clearly, it is imperative moving forward that the insurance and medical industry, as well as others responsible for erroneously labeling these accidents as "diving accidents," needs to stop. Categorizing or labeling these accidents into one group of "diving accidents" unfairly impacts the image of the sport of diving, but more importantly gives parents the false impression that these accidents are happening to trained and skilled athletes when they are not. It also falsely creates the general impression worldwide that the sport of competitive diving is unsafe when it is not. As a result of the labeling of these accidents and the stigma attached to the sport, many parents who should have their children take Learn to Dive lessons to help their child learn the basics of safe diving, just like they learn to swim, do not. Just as youngsters play soccer, T-ball, basketball, and flag football, they should participate in an introductory safe diving class so they, and their parents, know how to recognize dangerous situations in someone's backyard pool or an unfamiliar pool, and what to avoid in the natural environment. *Again, it is this expert's opinion that falsely labeling all of these accidents as "diving accidents" prevents parents from doing the one thing that they should do to possibly save their child's life or their ability to discern and survive a potentially traumatic accident—taking basic diving safety lessons.*

Moving forward, for public safety, I highly recommend that all entities responsible for categorizing accidents in the aquatic environment to consider classifying them correctly based on the accurate facts of each accident and reporting them accurately as:

1. *Recreational headfirst entry accidents*—Those accidents that occur in all types of pools and in the natural environment by novice, untrained recreational swimmers who are injured while going headfirst into a body of water.
2. *Sport of swimming headfirst entry accidents*—Those accidents that occur in practice, competition, or non-organized team activities in pools by a

trained swimmer injured while going headfirst, either from the starting block or the side of the pool, into a body of water.
3. *Sport of diving headfirst entry accidents*—Those accidents that occur in practice, competition, or non-organized team activities in pools by a trained diver injured while going headfirst, either from the board, platform, or the side of the pool, into a body of water.

Sadly, but truthfully, the first two categories will contain all the accidents, while hopefully the sport of diving will continue to not have any. As a parent myself, I want all parents to be armed with the facts. The fact is that the majority of these accidents happen to untrained recreational swimmers.

A NOTE FOR PARENTS

Educating and communicating with your children is the key. When interviewed afterward, almost all paralyzed victims said they had no idea they could get injured so severely. So explaining to your child that the water can be dangerous is a reasonable way to start the conversation. Telling them honestly that untrained, recreational swimmers can injure themselves in the pool or if they attempt to dive will bridge the gap to the lack of awareness that most people have.

Now it's a lot easier to have an impact on parents than to change public policy, property owners or agencies, or the natural environment. I am a parent too and fear is a great motivator. We want to protect our children at all costs. Parents should be concerned about their untrained child swimming or attempting to dive in the aquatic environment. It's a legitimate concern. As an aquatic expert, I highly recommend *enrolling your children in swimming and diving lessons* early on, no matter where you live. There are many attractions other than just the water that will get your children's attention and potentially harm them. Swimming, diving, and all the activities associated with the water are wonderful things for children to enjoy, but they are even more enjoyable when you and your children are armed with a healthy respect, knowledge, and education on how to swim and dive safely. We live in Florida, so our children grew up in the water—it's their playground all year around. In many other places that's not the case and children may, or may not, be in the water as much. The reality is that drowning among young children is a real threat and so is injury to the untrained swimmer who decides to take just one ill-advised dive. Please contact USA Diving, which is the national governing body for the sport of competitive diving and is recognized by the U.S. Olympic Committee at 317-237-5252 or online at www.usadiving.org.

As a parent myself, my top three concerns with children are

1. Shallow water.
2. Attending a friend's pool at their residence, without my own or competent adult supervision.
3. Going to a new pool environment for the very first time.

There are many other high-risk factors to be concerned about as a parent or anyone going into any type of aquatic environment, such as

- Alcohol. "Don't drink and dive." Nearly 50 percent of SCIs involve alcohol.
- Any type of headfirst diving in less than five feet of water.
- No lifeguard on duty.
- No depth marking.
- "No diving" markings.
- No warnings posted.
- No bottom markers.
- Poor lighting in and around the pool.
- Peer pressure.
- The temptations of the natural environment like a pier, rope swing, or bridge.

Things to remember and enforce with your children:

- "Feet first, first" whenever going into a pool.
- Look around and check the pool and pool bottom.
- Never dive the first time *anywhere*.
- Don't go headfirst down a slide or mat.
- Don't dive through inner tubes.
- Do not dive into above-ground pools.
- Don't pull or push anyone into the pool.
- Follow all rules at water parks and swimming pools.
- Follow posted signs in the natural environment.
- Use common sense.

IN CLOSING

The risk of severe injury and paralysis lies in the untrained, unsupervised, intoxicated negligence and bad judgment of swimmers entering shallow or unsafe unknown bodies of water headfirst. The responsibility also lies with people or entities associated with the design, planning, implementation, operation, or supervision of a facility where these injuries occur, and the regulatory agencies and individuals charged with responsibility to govern those facilities. Take personal responsibility for *your* safety and remember "feet first, first."

REFERENCE

Gabrielsen, M. A., McElhaney, J., and O'Brien, R. (2000). *Diving Injuries: Research Findings and Recommendations for Reducing Catastrophic Injuries*. Boca Raton, FL: CRC Press.

7 Ocean Beaches, Lakes, Quarries, Springs, and Canals

John R. Fletemeyer

CONTENTS

Ocean Beaches	138
Negligence Considerations	140
Inventory of Beach Hazards	140
Physical Human Hazards—Piers, Groins, Breakwaters, Beach Nourishment, Jetties	141
Jetties, Piers, and Groins	141
Groins, Breakwaters, and Jetties	142
Submerged Rocks and Boulders	143
Docks	143
Outfalls	144
Driving on the Beach	145
Dangerous Activities	146
Fires	148
Dogs	148
Beach Escarpments	149
Rip Currents	150
Diving into Shallow Water	152
Lightning	154
Water Pollution	154
Sharks	155
Jellyfish and Portuguese Man-o-War	156
Lion Fish	156
Stingrays	156
Sea Cliffs	157
Lakes	157
Canals	164
Quarries	165
Freshwater Springs	166
References	167

The objective of this chapter is to provide an overview of open water environments and the safety hazards that are associated with them. Failure to take measures and commonsense practices to respond to hazards and to protect the public often results in litigation alleging negligence and sometimes gross negligence if there are repetitions of the same accident.

OCEAN BEACHES

There are at least 12,000 miles of coastline in the United States and several thousands of beaches ranging from small pocket beaches to long, continuous, expansive beaches extending more than 100 miles. Beaches made up of boulders, shells, cobbles, mud, and sand and beach are significantly influenced by tides and the ocean bottom located immediately offshore, such as in coral reefs and sand bars.

Beaches are classified as being a high-, moderate-, or low-energy beach; and as being wave-dominated, tide-modified, or tide-dominated. Most California beaches are high-energy beaches and are wave-dominated by large waves while beaches on the west coast of Florida are mostly low-energy beaches and tide-modified with gradually sloping terraces and fine-grain sand. Beaches are composed of various materials besides sand including shell, boulders, cobbles, or mud. While the majority of American beaches are public, some are private, restricting or preventing public access. Laws in Hawaii require beaches to be accessible to the public regardless of private ownership. In some areas, although beaches are privately owned they must provide public access.

Beach recreation accounts for a multi-billion dollar industry and is responsible for thousands of jobs. In Florida it is estimated that beaches contributed more than $4 billion to the state's economy. Millions and sometimes billions of taxpayers' money are spent each year in this state on sand renourishment programs to replenish eroding beaches with sand and to maintain their recreational value. In 2012, more than $2 billion was spent renourishing beaches in Florida and New Jersey (Fletemeyer, 2017).

About 24 percent of the American population visits the beach at least once a year. This translates to about 75 million people visiting the beach. This number increases dramatically by several factors when considering that most people visit the beach multiple times a year. See Figure 7.1.

Beach recreation is one of America's favorite pastime activities, becoming popular at the turn of the nineteenth century. Ask any beach lifeguard and they will agree that beach safety, unlike swimming pool safety, represents special challenges because of the surprising number of hazards found in the coastal zone where beaches are located. It also represents a challenge because many hazards often appear suddenly and unexpectedly. This represents a significant difference than pool lifeguarding, where hazards associated with swimming pools are more predictable and consequently much easier and less problematic to respond to. In addition, there are significantly more hazards associated with beaches than pools. These hazards, when singular or combined, contribute to an unsafe beach requiring that stakeholders take appropriate action to resolve.

Ocean Beaches, Lakes, Quarries, Springs, and Canals

FIGURE 7.1 Beaches are magnets to the human population. An estimated 75 million people visit the beach in the United States every year.

Surf beaches have many hidden dangers that represent a threat to human safety, especially to the millions of bathers that visit the beach every year. Many of our beaches are protected by lifeguards, especially large sandy beaches located near metropolitan areas. Beach lifeguarding is much more challenging that guarding in pools. Consequently, more technical skills and a greater level of swimming ability are required. For this reason, individuals holding one of the several pool lifeguard certifications or the American Red Cross open certification are not recognized as being sufficiently trained or qualified to work on surf beaches. The United States Lifesaving Association (USLA) is responsible for training and certifying ocean lifeguards. Several lawsuits claiming lifeguard negligence have focused on the lack of adequate training and improper or out-of-date certifications.

Although not the primary focus of this book, this chapter would be remiss by not discussing beaches located in foreign countries. In the majority of cases, beach safety does not adhere to the same American standards and practices. Few beaches are guarded and for the ones that are, lifeguards are seldom adequately trained and do not have the arsenal of rescue and first-aid equipment typically found on American beaches. In addition, warnings about hazards are seldom provided (see Figure 7.2). In some countries, warning signs are vandalized or removed by the locals because of the negative publicity they are believed to generate.

Beach safety in third-world counties also represents a problem to the cruise ship industry because they often promote excursions to local beaches while docked in foreign ports. These cruise boat-sponsored excursions to local beaches are seldom evaluated by cruise ship staff and the cruise ship industry for safety. This failure has been responsible for several accidents and fatal drowning events.

FIGURE 7.2 Beach in Costa Rica where an American student drowned in a rip current. Note, there were no rip current warning signs posted on this beach.

NEGLIGENCE CONSIDERATIONS

To reiterate, the objective of this chapter and this book is to provide members of the legal profession an overview about beach and open water safety and to identify hazards that are most likely to be involved in a suit alleging negligence. In the majority of cases, lawsuits involving beaches focus on privately owned and supervised beaches. Publically owned beaches are often protected by "sovereign immunity." Consequently, they are seldom the subject of litigation because there is a limited opportunity for large recoveries because of "caps." However, there are exceptions to this because in some states recovery caps are large enough, making negligence suits more likely. In addition, in some extraordinary cases, legislative action is responsible for removing sovereign immunity cap recoveries.

It is important to note that most lawsuits alleging negligence often ignore accountability by the injured or deceased victim. As mentioned earlier, alcohol is involved in about 50 percent of fatal drowning events. Consequently, in a significant number of cases, the victim is often blamed for drinking and this may represent the proximate or at the very least a contributing cause of the victim's injury or death. Since drinking often plays a prominent role when defending a case, the results of a toxicology report becomes crucial.

Horseplay, risk-taking behavior, ignoring rule and hazard signs, and the lack of adult supervision represent other factors contributing to aquatic accidents and may also represent the proximate cause or a contributing cause of an accident. Often these factors are successfully used by defense attorneys to defend cases alleging either negligence or gross negligence. In rare cases, defense attorneys are able to win summary judgments citing the previously mentioned proximate and contributing causes.

INVENTORY OF BEACH HAZARDS

There is an amazing variety of beach hazards that are responsible for impacting public health and safety. Hazards can be classified as being either manmade or natural.

Examples of manmade hazards include piers, jetties, beach escarpments caused by beach nourishment, water pollution, and vehicles operating on the sand. Natural hazards include dangerous marine organisms (sharks, jellyfish, sting rays, red tide, and man-o-war), lightning, crashing and plunging waves, rip tides, and rip currents. Of all the natural hazards listed above, rip currents are hands-down the deadliest, accounting for at least 100 fatal drownings on American beaches every year.

Many beach hazards are ephemeral, meaning that they tend to appear suddenly and unexpectedly. Rip currents, a feeding shark, or an electrical storm can appear at a moment's notice and, therefore, demand an immediate response especially if the beach is protected by lifeguards. In most cases, this response involves either a verbal warning usually by a lifeguard or a symbolic warning strategically placed (i.e., a flag warning), or both. On some beaches without lifeguard coverage, law enforcement personnel assume the responsibility of warning the public.

In addition, many beach hazards may be relatively new to the environment, creating further beach safety challenges. Consequently, when a new hazard suddenly occurs, a new, creative, and decisive response is sometimes necessary. For example, recently there has been a proliferation of highly venomous lion fish in Florida and Bahamian waters. These colorful, diminutive fish were once native only to the South Pacific but when an aquarium owner released a few into warm Atlantic waters, they began to reproduce because they lack natural predators. Consequently, it has become necessary for local beach management staff to implement new warning programs devoted to these dangerous fish. NOAA and the Sea Grant program recently developed an education program in response to this finned and potentially dangerous hazard.

While some hazards can be neatly divided into either natural or human hazards, some belong to both categories. For example, shark fishing from the beach is a hazard that is both human and natural. At the time this book was being written in 2016, a spike in shark attacks occurred along the North and South Carolina, Florida, and California beaches. Some experts blame these attacks on an increase in shark fishing from the beach and "shark encounter" programs being promoted by the diving industry while others blame it on an increase in the population of prey species such as seals and sea otters.

In addition to the hazards discussed in the above, there are a host of hazards associated with recreational activities and human behavior. For example, surfing becomes a safety hazard when allowed to occur near bathers and if the consumption of alcohol is permitted on the beach, the opportunity for a serious accident or a drowning event is significantly increased. In some cases, if drinking is prohibited by law but not aggressively enforced, the consumption of alcohol may represent a serious hazard. There are many American beaches with laws prohibiting alcohol but where the prohibition is not actively enforced.

PHYSICAL HUMAN HAZARDS—PIERS, GROINS, BREAKWATERS, BEACH NOURISHMENT, JETTIES

Jetties, Piers, and Groins

Many beaches have fishing piers located within their boundaries. While piers are responsible for enhancing the beach environment, they also represent a serious safety hazard. Attached to pier pilings are sharp, razor-like barnacles and other harmful sessile organisms. These can cause severe lacerations, resulting in serious infections if not promptly treated. Discarded fishing hooks and monofilament line are other hazards associated with piers. In addition, fishing from piers is responsible for attracting predatory fish such as sharks, barracudas, and blue fish. In response to these hazards and to several others not mentioned in the above, municipal and private pier owners and operators have a duty to take proactive measures to protect bathers, surfers, and body boarders by keeping them away from piers.

In addition, there is always a danger of falling from a pier. This danger is increased at night. In response, many piers place lifesaving buoys within quick reach so that they can be tossed to a victim. Pier management should have an emergency action plan.

Groins, Breakwaters, and Jetties

Groins, breakwaters, and jetties (see Figure 7.3) have several functions including the following:

- Reducing wave energy
- Reducing beach erosion
- Providing entrances to navigation sea ports

Groins, breakwaters, and jetties are extremely hazardous, especially during times of heavy surf and when the tide is low. For this reason, measures must be taken to keep the bathing public away from these manmade structures. Usually the most effective measure is to provide warnings signs. When there are lifeguards nearby, lifeguards should provide verbal warnings and instructions to remain well away from these structures. Usually there is a long shore current and this current is often

FIGURE 7.3 Photo of a manmade jetty in Palm Beach. Rock jetties and breakwaters are common on many beaches. They are designed and engineered to reduce erosion and they are the source of many serious accidents.

Ocean Beaches, Lakes, Quarries, Springs, and Canals

responsible for drifting a victim into these structures. Consequently currents (speed and direction) must be carefully monitored and appropriate action taken. Failing to provide signage and verbal warnings resulting in an injury to a bather may represent "negligence" and a "professional breach of duty."

SUBMERGED ROCKS AND BOULDERS

Many natural beaches have submerged rocks and boulders. These natural features represent a hazard that may result in a serious head or spinal injury to a patron (see Figure 7.4). When rocks or boulders are present, warning signs should be posted. In some cases, small rocks have been removed from the ocean bottom.

DOCKS

Docks are manmade structures constructed from concrete, pressure-treated, or synthetic wood. Docks have several functions. Boat docks are the most common types of docks, ranging in size from small, private docks a few feet wide and a few feet long to mammoth-sized docks able to accommodate large ocean-going vessels. Docks are found in both fresh and salt water environments. This section applies to docks located in both.

Boat docks are not designed for public recreation but nevertheless may represent a hazard because they often are inviting. In the majority of cases, when someone sustains an injury from a boat dock, they are to blame. In some cases, no diving signs are placed at boat docks but this is an elective measure with no rules or regulations making this a requirement.

Docks are sometimes built for a recreational purpose or to define a bathing area (see Figure 7.5). Docks of this kind, because of their association with bathing, require measures preventing patrons from jumping or diving off. The most common safety measure is to post warning "No Diving" signs. The importance of warning signs is

FIGURE 7.4 Many beaches are dangerous because they have submerged rocks and boulders that can cause a serious head or spinal injury.

FIGURE 7.5 Docks constructed to define a bathing area or used for sunbathing invite jumping and diving. Warning signs must be posted to prevent this dangerous activity, especially if the dock is associated with shallow water. Nine feet is considered the minimum depth for safe diving.

increased if the dock is associated with shallow water less than nine feet. Nine feet is considered the minimum depth for safe diving but increases when diving from a high platform.

Docks are especially dangerous if located near bars and restaurants serving alcohol. This is a recipe for disaster because alcohol has been implicated in about 50 percent of serious accidents. Besides signs, bar staff must be trained and directed to warn patrons from to diving or jumping from a dock.

In addition to the dangers associated with diving, docks are also dangerous when located near bathing areas because of barnacles and other sessile organisms that are attached to pilings. These marine organisms can seriously lacerate a victim coming in contact with a dock piling. Besides signs warning bathers to stay away, pilings should be periodically maintained by scrapping off sharp barnacles with a wire brush or paint scraper.

Beside stationary docks, sometimes floating docks are placed offshore within a bathing area (see Figure 7.6). These often invite horseplay and diving, especially among teenagers. Consequently, floating docks must always be anchored in deep water with a nine-foot minimum depth according to FINA standards and should be regularly maintained.

OUTFALLS

Some beaches have drainage outfalls extending across the beach. Sometimes outfalls are exposed, while others are buried under the sand. Outfalls function to transport greywater and rainwater offshore. If the outfall is exposed and not buried, erosion gullies often form, creating a hazardous condition, especially to children (see Figure 7.7). Stakeholders owning or controlling the beach must take measures to minimize

Ocean Beaches, Lakes, Quarries, Springs, and Canals

FIGURE 7.6 Floating docks designed for recreation should be anchored in deep water to prevent a diving injury.

FIGURE 7.7 Open outfall stretching across Panama City Beach. Outfalls are often the source of pollution and consequently represent a hazard, especially to children.

this hazard. Regarding open and exposed outfalls, these should be eliminated especially if located on a recreational beach.

DRIVING ON THE BEACH

Several serious accidents, including some fatal, have resulted when either a police, service, or lifeguard vehicle struck a pedestrian. Since 1995, Volusia County, Florida, beaches have had 49 beach-driving accidents in which a pedestrian was

FIGURE 7.8 Driving on the beach is a serious hazard because rules intended for the street don't apply and because bathers often share the area where vehicles are driven.

either injured or killed. In 2013, there were two four-year-olds who were killed in separate accidents. On a state-owned park beach in Southeast Florida, a toddler was driven over and killed by a park ranger driving an ATV. A host of other beaches have had serious or fatal accidents resulting from negligent and reckless vehicle operation (see Figure 7.8).

It is incumbent that strict guidelines are established and enforced to prevent beach accidents involving vehicles. Following a sunbather being run over by a lifeguard jeep on a beach in Florida, a policy was established requiring lifeguard vehicles to be driven only parallel to the ocean and during times that the beach is not crowded. In addition, following this accident, warning beepers were installed on every lifeguard vehicle.

On a Volusia County beach, after lifeguards ran over three bathers over a twelve-month period, all lifeguards are now required to take a driving course identifying the importance of identifying blind spots and watching out for sunbathers. At the state park beach involving the ATV, park rangers must now operate their vehicles at slow speeds only, even when responding to an emergency. In Palm Beach, even without a history of pedestrian accidents, lifeguards are required to complete an ATV safe driving course administered by the police department before operating an ATV on the beach. Considering the several noted examples, with any accident occurring on the beach, whether involving a public service vehicle or a privately operated vehicle, operator negligence is likely the cause. This has been established by case law involving law enforcement personnel responding to a land-based emergency and resulting in a civilian injury while in reroute.

Dangerous Activities

On many beaches, recreational activities occur that represent a hazard to bathers. The activities listed below should not be allowed in the vicinity of bathers. When

Ocean Beaches, Lakes, Quarries, Springs, and Canals

accidents occur, negligence is often related to a stakeholder's failure to identify and enforce preventative measures, including warning signs. In legal terms, Mr. John Doe knew or should have known that it is not safe to allow surfing in an area designated for swimming. The following is a list of potential hazards where there is a duty for stakeholders to respond:

- Fishing. Fishing should never be allowed in a bathing area. Discarded fishing hooks create a safety hazard. Bait in the water often attracts large predatory fish, such as sharks.
- Surfing. Surfing and bathing should be clearly separated into defined and marked areas. Accidents occur when a runaway surf boat strikes a pedestrian in the head or face.
- Kayaking, surf skiing, and windsurfing. Kayaks, surf skies, and windsurfing boards can strike and injure victims if not properly controlled by the user. For this reason, it is incumbent for stakeholders to clearly identify activity areas. In many cases, water avenues are created on the beach, allowing kayakers and windsurfers to safely navigate in and out of the beach. Once beyond the bathing, boaters are free to operate without restrictions.
- Power boating. Power boats must never be allowed in an area where there are bathers. For this reason, most beaches prohibit power boats from operating within a defined bathing area. Typically a restricted area between swimming and boating should be at least 100 feet, and in some cases, 300 feet.
- Kite flying. Accidents are fairly common, especially if kites are launched from the beach. Lines attached to kites have been responsible for injury to a bather walking on the sand. Anytime that a member of the bathing public is injured by kite, the operator should be held responsible and negligent.
- Ball and Frisbee play. While the number is not known, bathers and beachgoers are frequently injured by a Frisbee or hardball, such as a football. Eye and facial injuries are most common and on a crowded beach is when this type of injury is most likely to occur. On any beach promoting public recreation, ball and Frisbee play should be restricted completely or at the very least, restricted to a designated area.
- Diving from elevated structures, that is, bridges, piers, and cliffs. Any elevated structure is a potential hazard, especially to young males who often engage in risk-taking behavior. When possible, warnings should be provided, and if the hazard is considered significant, more aggressive measures should be taken, including enforcement by public safety officers.
- Snorkeling. In the past, snorkeling was considered a relatively safe activity. However, over the last decade, a condition called "shallow water blackout" has been implicated in more and more drowning deaths (see Figure 7.9). The complex physiology of this condition is discussed in detail in Chapter 2. In the majority of cases, shallow water blackout occurs when an individual hyperventilates several times before going underwater. Because snorkeling is an activity that often involves a series of hyperventilations, this activity may be especially vulnerable to this condition. On beaches that are

FIGURE 7.9 Snorkeling is a hazardous activity because of a condition called "shallow water blackout."

lifeguarded, guards should warn snorkelers about this hazard and should always be especially vigilant when a bather is engaged in this breath-holding activity.

FIRES

Fires create a safety hazard for several reasons. After fires are extinguished, the hot coals and wood residue can injury bathers sometimes long after the fire has been extinguished. Fires also impact wildlife, especially hatchling turtles, by making them disoriented by the light originating from the fire. Beach pollution is another result caused by fires on the beach.

DOGS

Many beaches allow dogs but only in certain designated areas. This is often controversial, especially if dogs are permitted to be unleashed. A dog bite may represent a liability exposure not only to the owner but also to the entity permitting this practice. In addition, dogs are sometimes responsible for impacting the coastal zone. For example, unleashed dogs may impact endangered or threatened turtles and may chase sea birds from their nests. Least terns are small sea birds that often nest on isolated sections of the beach and may abandon their nests if threatened by a barking or chasing dog. In addition, since most beaches are sand beaches, dog excrement is nearly impossible to remove from the sand. This creates health hazards to children building sandcastles or simply playing in the sand. In response to these issues, some municipalities allow dogs on the beach but only at specified times, usually during the late afternoon or early evening. See Figure 7.10.

FIGURE 7.10 Unleashed dogs represent a safety hazard (i.e., dog bites) and health hazard because dog excrement cannot be removed from the sand.

BEACH ESCARPMENTS

Beach escarpments with high, vertical profiles are usually located near the high tide line. They can form naturally or form as a result of human intervention, including beach sand nourishment programs. In some cases, escarpments can more than six feet high and nearly vertical (see Figure 7.11). Escarpments can cause a serious injury in the event that a bather falls from one. They are especially dangerous at night when a lack of lighting prevents them from being seen and avoided. In some cases when escarpments occur on lifeguarded, public beaches, they are responsible for creating a visual barrier and preventing lifeguards from seeing bathers located immediately offshore. Escarpment hazards should be responded to by providing warnings (see Figure 7.12) or by a removal program involving mechanical equipment that assists in grading and leveling the escarpment.

FIGURE 7.11 Escarpments create a hazardous beach condition. Falling from a high-profile escarpment such as the one in the photo can result in a broken bone or spinal injury.

FIGURE 7.12 Signs are often placed to warn bathers about escarpments (cliffs).

Rip Currents

Of all beach hazards, rip currents are the deadliest. Haus discusses rip currents from a different perspective. Rip currents are responsible for at least 100 fatal drowning fatalities every year (see Figure 7.13). Rip currents are found on most salt water beaches and even on a few freshwater beaches, especially Lake Michigan, where in 2011 nine drowning deaths were attributed to rip currents.

Rip currents are not to be confused with rip tides and undertows. The media commonly makes this mistake. Consequently, they incorrectly warn about the presence of rip tides instead of rip currents. Rip currents are strong, usually temporary currents that flow in a direction away from the beach. The classical description of a rip

Ocean Beaches, Lakes, Quarries, Springs, and Canals

FIGURE 7.13 Rip currents are found on most ocean beaches and account for more than 100 fatalities on American beaches every year. Note that the photo of this rip current does not conform to the typical rip current description.

current was made by Kummer (1952). He describes a rip current as having a slender neck terminating offshore in a shape resembling the head of a mushroom. While this description is fairly accurate, rip currents often are shaped differently than the classical description:

> Research focusing on these deadly currents reports that there are two types of rip currents. One is a "permanent" rip current that is often found on a pocket beach, near a pier or near a natural rock outcrop. The second type is a "temporary" rip. Some rip currents are called "flash" rip currents. Flash rip currents appear quickly and then quickly disappear. Flash rips can be particularly dangerous because they often catch lifeguards and public by surprise. Some rip currents migrate down the beach usually in the flow direction of the littoral of longshore drift located just offshore of the beach.

In recent years, research has made several new and important discoveries about rip currents. For example, research has found that rip currents often demonstrate a "pulsing" behavior. As the rip current begins to pulse, its strength intensifies, suddenly making it significantly more dangerous to swimmers. NOAA reports that the most dangerous rip currents occur during outgoing tides and during low tide. While rip currents can be expected on any ocean beach, they are especially prevalent on high-energy beaches, pocket beaches, and beaches where a system of sand bars is located immediately offshore (see Figure 7.14).

Also, recent research by Fletemeyer (2012) discovered that rip currents are exceedingly difficult to observe while standing on the beach and looking for some of the characteristics believed to be common among all rip currents, including trapped debris that is moving quickly away from the beach, choppy or excited water located in the "neck" of the rip current, and a discoloration in the water between the water located within the rip current and the surrounding water. He reported that these "visual" characteristics may not be reliable indicators and the belief that the public can be trained to spot rip currents is not realistic. In addition, even experienced

FIGURE 7.14 Sand bars located immediately offshore are often responsible for creating strong rip currents, especially on high-energy beaches. Photo shows sand bars on Panama City beach where rip currents account for several fatal drownings every year.

lifeguards may not always be able to spot rip currents from the beach. Consequently, is necessary for lifeguards to periodically enter the water and to "feel" for the current generated by the rip.

Research suggests that the rip current warnings that proliferate on our beaches are not working as intended. Flag system warnings are especially ineffective because flags represent a symbol and for a bather to understand and respond to the symbol, they must have additional information explaining the meaning of the symbol. In a study about beach warning sign effectiveness, Fletemeyer (1999) reported that a significant number of bathers (especially tourists) are not able to accurately describe the difference between a yellow and red flag and were confused about the intended message for a red flag and a double red flag. In addition, it was reported that traditional messages about how to escape a rip current may not be accurate on some beaches where there is a strong longshore current, usually flowing in the same direction. There are many cases that have been filed and settled alleging that a bather was not properly warned when rip current conditions were known to exist.

It has been reported that rip current drowning events are most likely to occur on sunny days and when the surf conditions are moderate. This may seem counterintuitive but in reality it is not. In very rough water and during inclement weather when rip currents can always be expected, conditions are such that most bathers do not come to the beach. When the surf is rough, these conditions are sufficient to keep bathers with marginal swimming skills out of the water.

DIVING INTO SHALLOW WATER

One of the most serious and insidious injuries that occur on beaches are spinal injuries (see Figure 7.15). The record is replete with primarily teenage boys and young men "crashing" through the surf while diving head first into the ocean bottom. In some cases, large "crashing" and "spilling" will "pile drive" a victim into the sea

FIGURE 7.15 Diving into shallow water ranks as one of the most significant beach hazards. Many beaches especially located next to hotels and resorts post warning signs encouraging bathers not to dive into shallow water.

bottom, causing a serious and life-threatening head and/or spinal injury. This may result in quadriplegia or even death.

A study of spinal injury conducted by the Hoag Hospital in California found that about 80 percent all serious head and spinal injuries occurred among males engaged in risky behavior. An unpublished study conducted by Fletemeyer (1993) reported that when in the surf zone and confronted with a large wave, males usually attempt to dive through or under the wave while females typical jumped up, trying to go over the wave. This difference in behavior may account for the huge disparity in spinal injury between males and females.

Emergency response by a lifeguard or first responder (EMT or paramedic) to a spinal injury in the surf requires special training and usually requires "packaging" the victim in the water unless the victim is not breathing and does not have a pulse. In addition, if the water is rough, then sometimes packing a victim in the water is not possible. Failure of proper spinal management by the first responder may seriously exacerbate the injury. If the rescuer is a lifeguard or a professional first responder, there may be serious liability exposure if it can be proven that the victim was improperly treated. In addition, it is crucial to have a c-spine board readily available to safely and successfully transport the injured victim out of the water and onto the beach. There have been several cases where a c-spine board was locked in a room and consequently several crucial minutes were wasted locating it.

On private beaches controlled by hotels and resorts and where guests have sustained a spinal injury, hotel staff have responded without being properly trained. Consequently, they have been responsible for making the victim's injury worse. Hotels on the beach must provide staff trained in spinal injury management and emergency first aid.

On crowded beaches, shallow water "No Diving" signs are sometimes posted. It should be noted that alcohol is often involved in shallow water diving accidents, which is all the more reason to establish and enforce rules that keep alcohol off the beach.

LIGHTNING

Lighting is a major hazard, threatening public safety not only in pools as discussed in Chapter 2 but also on the beach. On beaches that are guarded, lifeguards have a duty to quickly respond to an electrical storm by clearing pedestrians and bathers from the beach. This should be established in an SOP manual.

In the United States, an average of 62 people a year are killed by lightning. Hundreds of others are injured. Because of the openness of the beach, bathers are particularly vulnerable to being struck by lightning. As a rule of thumb, the National Oceanic and Atmospheric Administration (NOAA) recommends waiting 30 minutes after the last thunder crack before going back to the beach. This recommendation should also be followed by lifeguards and adopted as a policy.

WATER POLLUTION

Water pollution is a serious hazard that can cause illness and even death, especially among the young and elderly. Gastroenteritis (diarrhea and vomiting) is the most common waterborne illness related to pollution. Annually, several thousand closures occur on American beaches. In 2002, the CDC reported that beach pollution prompted 12,000 closing and swimming advisories at ocean, bay, Great Lake, and some freshwater beaches. About 87 percent of the beach closures were issued after the presence of bacteria associated with fecal contamination was detected. See Figure 7.16.

Too many nutrients can cause harmful algal blooms and fish kills. Excessive nutrients can come from sewage treatment plants, boating wastes (illegal bilge pumping), industrial discharge, and deposition from air and ground water contamination. Sewage overflows can release pathogens or disease-causing microorganisms. Pathogens can cause a wide range of health problems, from sore throats to meningitis.

FIGURE 7.16 Water quality must be continuously monitored and when water becomes contaminated for any reason, signs must be posted and measures taken to prevent swimming and engaging in aquatic activities.

In addition, a phenomena called "red tide" may cause toxicity in shellfish, making them poisonous to eat. Typically when red tides occur, government agencies post warnings and advisories not to harvest shellfish at certain times. When beach pollution is identified, stakeholders have a duty to provide the public warnings until the pollution declines to an acceptable limit. Failure to provide warnings may represent a claim of negligence.

Sharks

The threat of a shark attack has been exaggerated, especially by the media. In the United States, shark attacks are exceedingly rare events, although in the year 2015 during the late spring and early summer there was a spike in shark attacks, especially along North Carolina beaches. In addition, more and more sharks are being spotted cruising close to shore. This may be the consequence of an increase in the popularity of surf fishing and using bloody bait.

In addition, shark encounter programs have gained in popularity and occur in coastal waters not only in the United States, but abroad where shark populations are abundant. Anytime humans associate with sharks, there is an opportunity for an unprovoked attack. Feeding sharks creates a heightened risk because this may result in initiating "frenzy" behavior, making humans a prey target. Consequently, considerable caution must be taken and feeding should never be allowed. Water clarity must always be optimal and the dive leader (usually a certified dive master) must always have control while assuming responsibility for the safety of the diver whether novice or experienced. In the event that a particularly aggressive species of shark is encountered (i.e., great white, sand tiger, bull, or dusky), divers should be removed immediately from the water.

In most cases, shark attacks occur inshore of a sandbar or between sandbars where sharks become trapped by low tide. Shark attacks are more likely to occur in the early morning and at night and during times when the water is murky. In regard to murky water, a shark is more likely to mistake a human for prey. Sharks are more common during certain times of the year. In Florida, large congregations of feeding and migrating sharks occur during the early fall months in response to the mullet runs.

As in the case of any known hazard that threatens public safety, it is incumbent to warn the public. Failure to provide appropriate and timely warnings often results of allegations of negligence when a bather is injured or killed.

When sharks are observed or even suspected, warnings must be posted. On beaches serviced by lifeguards, lifeguards must be trained to be diligent for sharks. When observed, the water must be cleared until it can be certain that the shark(s) is no longer in the immediate area.

To reduce the risk of a shark attack, NOAA makes the following recommendations:

- Don't swim too far from shore.
- Stay in groups—sharks are more likely to attack a solitary individual.
- Avoid being in the water during darkness or twilight when sharks are most active.

- Don't go in the water bleeding from a wound.
- Don't wear shiny jewelry in the water.
- Don't swim in the water when migrating and feeding sharks are reported.
- Never swim in an area where people are fishing.

Regarding fishing, many beaches have enacted ordinances prohibiting shark fishing from the beach. Targeting sharks by surf fisherman is a recipe for a possible shark attack.

Jellyfish and Portuguese Man-o-War

Most jellyfish and all man-o-war have microscopic stinging cells on their tentacles called *nematocysts*. A sting from a jellyfish or a man-o-war may result in a temporary welt on the extremity after having physical contact with the organism, or in extreme cases, may result in death. In cases involving death, the victim is usually allergic to bee stings and quickly suffers a condition called *anaphylactic shock*. Victims suffering from a jellyfish sting should be immediately treated by first removing the tentacle. Sometimes sand is rubbed into the skin but this should never be done because it causes the poison from the nematocysts to go deeper in the body. If the victim's condition worsens, medical help should be summoned immediately.

On many beaches the presence of man-o-war is seasonal. On Florida beaches, January and February are often regarded as man-o-season. When man-o-war is spotted on the beach, lifeguards have a duty to warn the public (see Figure 7.17). In rare cases, a bather stung by a man-o-war suffers a condition of anaphylactic shock. This condition may require emergency room intervention and if not immediately treated may result in death.

Lion Fish

Lion fish were native to the South Pacific. However, when these fish were inadvertently released into tropical Atlantic waters, they began to proliferate due to a lack of natural predators. These fish inject a neurotoxin from their dorsal fins, causing the victim to suffer excruciating pain. In rare cases, death may occur, especially among the young and elderly. Lion fish are especially dangerous because they are often found in shallow-water bathing areas or around jetties where bathers like to snorkel. In areas where lion fish are known to exist, warnings should be posted.

Stingrays

Stingrays are a member of the shark family and are commonly found in shallow water, especially during the summer time. Consequently, many stingray injuries occur when they are accidentally stepped on. In most cases, when stepped on, the stingray will defend itself by injecting a serrated barb into the foot or heel area. If this should happen, medical help immediately should be summoned. A barb should never be removed by a person without emergency first aid training.

Ocean Beaches, Lakes, Quarries, Springs, and Canals

FIGURE 7.17   Man-o-war warning posted a Florida beach.

Sea Cliffs

Many beaches are located near sea cliffs. Hazardous cliffs are common in certain areas of California, Oregon, and Hawaii. Cliffs represent a significant hazard and warnings should be posted when sea cliffs are located by beaches that are heavily populated by bathers and fisherman.

Lakes

In addition to ocean beaches, freshwater lakes provide recreational opportunities to countless millions, especially during the summer months when the weather is good and school is out (see Figure 7.18). There are two categories of lakes, manmade and natural. Lakes number in the thousands. For good reason, Minnesota is called the

FIGURE 7.18 Freshwater lakes provide recreational opportunities to millions, especially during the summer months. Some lakes provided designated bathing areas that must adhere to several safety standards and practices.

land of 10,000 lakes. Lakes vary in size from only a couple of acres (called ponds) to the massive size of Lake Michigan, which has a surface area of over 24,000 square miles. As in the case of ocean beaches, freshwater lakes have a surprising variety of hazards that threaten public safety. Take for example rip currents, which were believed to be only an ocean hazard. In 2010, eleven bathers drowned after being caught in rip currents in Lake Michigan.

Lakes have an enormous recreational value. While accurate data about this subject is lacking, Keshlear (1999), an employee of the U.S. Army Corps of Engineers, reported that in 1995, lakes managed by the Army Corps registered more than 385 million visits by the public. These visits had an economic impact of over 20 billion dollars and were responsible for creating over 600,000 full-time and part-time jobs.

Lakes, whether natural or manmade, are viewed by the courts as "attractive nuisances." Consequently, this concept is sometimes cited in complaints as being a contributing cause to a serious accident or fatal drowning (see Figure 7.19).

Recreational lakes are often the sources of serious water accidents and fatal drowning events. Consequently, suits alleging negligence and gross negligence result from the lack of proper warnings. In these cases, even when there are appropriate warnings, several mitigating issues are sometimes involved (i.e., the consumption of alcohol resulting from a positive toxicology report).

Lakes that are intended and promoted for bathing should have the following characteristics related to safety:

- A defined and well-marked designated bathing area. Usually designated by an anchored buoy and line system.
- A designated 911 phone, especially if cell phone coverage is an issue.
- Free of sudden transitions from shallow water to deep water. No deep holes or water pockets.

Ocean Beaches, Lakes, Quarries, Springs, and Canals

FIGURE 7.19 Ponds and small lakes are dangerous because they are considered attractive nuisances and because they are often located in urbanized areas where children play. This photo shows a small lake/pond located next to a housing development where a young boy fatally drowned.

- Identification of or measures taken including posted warnings when dangerous animals are known to inhabit the lake, for example, alligators (see Figure 7.20).
- Free of entangling debris and vegetation (see Figure 7.21).

FIGURE 7.20 Like ocean beaches, freshwater environments, including lakes have many hidden hazards that threaten public safety. Signs must be conspicuously posted warning about these hazards.

FIGURE 7.21 In this photo of a drained lake underwater debris is exposed demonstrating a hidden danger that could entangle and drown a bather.

- Lifesaving rescue buoys strategically placed along the shoreline (see Figure 7.22).
- Proximity to an automated external defibrillator (AED).
- Staff and stakeholders trained to respond to a drowning event.

Warning signs represent the first step in warning guests and bathers about lake hazards. Failure to provide warnings may be either the proximate cause or a contributing factor in a drowning event. When providing warnings, careful planning is required because warnings are often not effective unless they are strategically and conspicuously located for all to see. In addition, they should confirm to ANSI standards.

An example of freshwater lakes without proper warnings is provided by Roger Strassburg in Chapter 11. Strassburg discusses the impact of a brain-eating ameba (*Nigeri fowleri*), a microscopic organism that has been responsible for killing more than 100 individuals in lakes primarily located in Minnesota, Florida, and Texas. This attorney argues about the duty for the lake stakeholders to provide warnings in lakes where these deadly organisms are known to thrive.

While there are no set standards about how lakes should be designed for recreation, the U.S. Army Corps has made several important recommendations related to engineering and design issues (see Keshlear, 1999).

Recreation lakes designed for bathing should be engineered to have a gradual slope leading from the shore to deeper water and should be absent of any sudden drop-offs.

The slope of the land, both above and below the waterline, is one of the determining factors in the selection of a good beach site. Slopes in the underwater portion of beaches should ideally range from 2 to 5 percent, with the most desirable slope being as flat as possible to disperse use. Beach bottoms are designed to eliminate sudden changes in grade or drop-offs in the 0–5 foot depth. Pre- and post-impoundment

Ocean Beaches, Lakes, Quarries, Springs, and Canals

FIGURE 7.22 Lakes promoting bathing and without lifeguards should strategically place lifesaving rings on the shoreline. Lifesaving rings have been responsible for rescuing drowning victims by members of the public without lifesaving skills.

studies are performed to ensure acceptability of gradients. Daily, seasonal, and yearly water level fluctuations due to irrigation, flood control, evaporation, power generation, or other factors are considered in beach design to ensure optimum utilization. A detailed in-section of the underwater portion of the beach is accomplished each year just prior to opening to the public. The inspection includes the necessary detail to reveal sinkholes, depressions, or dangerous drift material, and corrective actions are taken prior to the opening of the beach. Inspection records are maintained on file.

Water quality is an important consideration when beaches are involved and may represent a contributing cause to a serious accident or fatality. Clarity (turbidity) can be evaluated two ways, using a Secchi disk or by using a ratio turbidmeter. Water quality at all beach locations must be acceptable for swimming. Prior to detailed design, water quality sampling data is collected, analyzed, and coordinated with appropriate state agencies.

In addition, lake bottoms should be free of debris such as tree stumps or entangling vegetation. Regarding vegetation, it may necessary to conduct an aquatic weed control program. In addition, recreational lakes should never allow power boating to occur in or near designated swimming areas. To avoid this problem, stakeholders have enacted regulations allowing only non-power boats on lakes, or in some cases, allowing only power boats equipped with electric motors.

Some recreational lakes provide summer lifeguards. While lifeguarded lakes are significantly safer than lakes without lifeguards (Branche, Fletemeyer et al., 2001), the practice of providing lifeguards presents several challenges and if not met properly, creates a significant liability exposure. Lake lifeguards must be open water certified by the American Red Cross. Lifeguards having only swimming pool certifications are not properly certified and therefore not trained to guard lakes or other open water environments.

In addition, stakeholders are responsible for seeing that there are an adequate number of lifeguards given the size of the area that they are being required to guard. There must be a sufficient number of lifeguards, allowing them to take periodic breaks, to eat lunch, and to take into account unplanned absences due to illness. In addition, lifeguards should be provided with umbrellas or towers, allowing adequate shade and ventilation. Generally, elevated towers and lifeguard stands should be strategically placed where lifeguards are not forced to glare directly into the sun when watching the water. Glare from an early morning or later afternoon sun may prevent a guard from seeing a victim in distress or in the early stages of an active drowning event.

Open water lifeguards must be provided with certain basic equipment that is designed to facilitate victim rescue, including, buoys, swimming flippers, c-spine boards, and rescue boards. In addition, there should be at least one AED, and all lifeguards should be equipped with first aid and personal safety items associated with CPR and bloodborne disease prevention, including barrier masks, gloves, gowns, and Ambu bags.

Lifeguard staff should be given recurrent training and sufficient time to train and to practice rescue skills learned in class. This is often referred to as a "recurrent" training program and is often a subject in a lawsuit, although seldom identified as being the proximate cause. Recurrent training activities should be logged in a book and may be useful when defending a suit alleging lifeguard negligence.

Non-guarded lakes, whether designed for recreation or not, must provide warnings related to the hazards that are known to exist (Figure 7.23). It is necessary for stakeholders to conduct a comprehensive risk management survey that is designed to identify and to list every life-threatening hazard (refer to Chapter 5). In a majority of cases, this task requires the help of a professional consultant or in some cases, a team of consultants. Once hazards are identified, there are two industry-accepted responses:

- Through aggressive action removing or eliminating the hazard, for example, removing stumps from the lake bottom, dredging the lake to eliminate sudden drop-offs, or trapping and removal of dangerous aquatic life.

Ocean Beaches, Lakes, Quarries, Springs, and Canals

FIGURE 7.23 Freshwater lakes have main hazards that require warnings. In this photo a warning is provided about jumping from a bridge extending across a lake.

- If removing the hazard is not possible, then provide appropriate warnings. In most cases, these involve written warnings that conform to ANSI standards requiring signs and letters to be of a specific shape, size, and color. Fletemeyer (2001) reports that written warnings provide two vital functions—they warn the public about a hazard that was not previously known to exist or they serve as reminders about a hazard that is known to exist. Regarding warnings, they must be periodically maintained. Warning signs are often vandalized and ultraviolet radiation from the sun and weathering may cause letters to fade, making signs difficult to read.

If lakes do not have the benefit of lifeguard protection, police officers and other emergency personnel should be provided water safety training (Fletemeyer 1999). Every officer and paramedic with a lake, canal, or river located within their jurisdiction should know how to effect a rescue without putting their own lives in danger.

CANALS

Canals are especially dangerous because they often have steep and slippery embankments leading directly to deep water (see Figure 7.24). The All-American Canal is 60 miles long and is located east of San Diego. This canal had more than 500 documented fatal drownings over a ten-year period. However, thanks to some design changes, including posting warning signs and safety ropes every 100 meters across the canal, the loss of life was dramatically reduced. In two years following the improvements, only one drowning has been reported.

If canals are located in urban areas or near school campuses, they should be fenced. In addition, warnings should be posted (see Figure 7.25), and in some cases, dedicated lifesaving rings should be established in strategic areas. If stakeholders become aware of a problem, ignoring it may have legal and liability consequences. For example, if a school administrator is aware that students are sometimes "sneaking" into a nearby canal to swim, the administrator has a duty to respond even when occurring off campus. At the very least, administration has a duty to notify law enforcement personnel and work cooperatively to eliminate the problem.

To reiterate, nearly every open water environment represents what is called in legal terms an "attractive nuisance." Canals are especially dangerous because they

FIGURE 7.24 Typical canal with steep embankment leading directly to deep water.

Ocean Beaches, Lakes, Quarries, Springs, and Canals

FIGURE 7.25   A warning sign posted next to a canal embankment warning of deep water.

are often located near urban areas. Consequently, it is incumbent for stake holders, whether representing the public or private sector, to take an active and aggressive role in preventing accidents.

QUARRIES

Quarries are inherently dangerous because they were never designed for swimming or bathing. Most quarries have steep rock faces leading directly into deep water. In addition, some quarries have high cliff faces that attract risk-taking behavior primarily among teenage boys. Diving from high quarry cliffs is often a rite of passage for teens.

A majority of quarries have been abandoned. Consequently, trespassing by young males between 10 and 20 years old accounts for most quarry deaths. It has been estimated that quarries account for between 20 and 30 fatalities per year and that a majority of these fatalities are drowning.

A few quarries that are privately owned have been converted into campgrounds offering a variety of aquatic activities including bathing, fishing, small craft boating (canoeing), and SCUBA diving. Because quarries are often are spring fed, they have exceptionally clear water, and consequently, are often popular places for SCUBA divers. There have been several SCUBA quarry drownings, mostly involving novice divers with little or no open water diving experience.

For a number of reasons, quarries are dangerous and have a heightened level of liability when compared to other freshwater aquatic environments. Some of the dangers often associated with quarries include the following:

- Because quarries are deep and usually spring fed, they often have a thermocline with an abrupt transition between warmer and much colder water. Upon reaching a thermocline, a bather or inexperienced SCUBA diver might begin to hyperventilate and panic. Encountering a thermocline is especially dangerous to the elderly or to a person suffering from asthma. It has been documented that cold water can sometimes trigger an asthma attack.
- In the case when quarries are abandoned and not designated for recreation, trespassers, especially teenagers, may illegally venture into a quarry and not be aware of some of the hazards referenced above. When engaged in trespassing, the ownership of a drowning or accident is the trespasser and not the owner.
- In the case of quarries that are used for recreation, developing rules and enforcing them is often problematic for the owner and or staff.
- People visiting quarries often ignore safety rules, especially rules prohibiting alcohol consumption.
- Because quarries were never designed for recreation, providing a safe bathing environment is often challenging. Perhaps the biggest challenges relates to the deep water and the transitional bottom topography. Frequently docks with ladders are built to respond to this inherent condition.
- Water conditions are constantly changing, including water temperature and visibility during a rainy day, water visibility my drop significantly.
- Quarries may have dangerous animals inhabiting the premises, such as poisonous snakes.

Unlike swimming pools, quarries are not stringently regulated by codes and regulations. Quarries offering bathing and other recreational opportunities are not required to have lifeguards and consequently, most do not. In most cases when quarries allow swimming, patrons are instructed that swimming is conducted "at your own risk." Usually "swim at your own risk" warnings are accomplished by both written and verbal instruction.

While quarries have a long history of negligence suits, placing the blame often becomes a double-edged sword with the victim sometimes owning much if not all of the blame. Horseplay, risk-taking behavior, ignoring posted rules, trespassing, and the consumption of alcohol are often contributing factors when a bather is either injured or drowns in a quarry.

Freshwater Springs

Like rock quarries, freshwater springs were never designed for bathing. Consequently, like quarries, freshwater springs often have sudden transitions leading from shallow to deep water. At some freshwater springs located in northern Florida, the water depth exceeds 100 feet. In addition, many freshwater springs have strong and unpredictable currents that often add to their danger, especially to non-swimmers and among swimmers with limited swimming ability.

Because many springs have strong currents and often have exceptionally deep water, it is incumbent for owners to have effective warning signs that are strategically located for patrons to see and that comply with ANSI standards. In addition, because many freshwater springs are popular venues for SCUBA diving, SCUBA diving must be carefully regulated.

Although relatively few in number when compared to lakes, quarries and freshwater springs are popular bathing areas. The majority of these are privately owned but a few are owned by the federal, state, or local government. Quarries and freshwater springs are dangerous for several reasons. First and foremost, being natural bodies of water, they were never designed for bathing. In addition, most are deep with sudden transitions between shallow and to deep water. In the case of some freshwater springs, the water quickly transitions from only a few feet to more than 100 feet. In most cases, these two aquatic environments have steep embankments, making entry and exit difficult, especially for the old and young. In the case of the freshwater springs, many have strong, localized currents.

Freshwater springs offer an inviting place for bathing and swimming for several reasons, including exceptional water clarity and an abundance of aquatic life. Several springs in Florida often attract manatees. Unlike swimming pools, they are not regulated by codes or government regulations. Consequently, being self-regulated, many lack appropriate warnings and they are often inadequately supervised with few if any safety rules and regulations. The consumption of alcohol is often openly conducted and these aquatic environments are usually located a considerable distance from emergency help. Consequently, when an emergency occurs, it may take considerable time for EMS or paramedic personnel to arrive on the scene. It is, therefore, incumbent for owners and staff to have emergency first aid training and to have emergency aid equipment including an AED onsite. An inspection of many of these sites conducted by the author found that none had an AED. Based on these issues, freshwater springs have significant liability exposure and are often the source of a water accident or drowning event. In the majority of cases of a serious accident or drowning, they would have been preventable had owners and staff taken reasonable accident prevention measures and practices. However, in some cases involving trespassing, this is not the case.

REFERENCES

ANSI regulations for warning signs (2006). *Specification for Preventive Signs and Tags.* 1910.145, Occupational and Health Standards.

Branche, C., and Fletemeyer, J. (2001). Lifeguard effectiveness: A report of the working group. Centers for Disease Control Spec. Publication, pp. 1–36.

Fletemeyer, J. (1999). Warning sign effectiveness. Spec. report funded by Panama City Beach Dept. of Tourism, pp. 1–44.

Fletemeyer, J. (2001). All-American canal safety measures. Report and published by the San Diego Water Authority and the Colorado River Authority, pp. 1–61.

Fletemeyer, J. (2008). Rip currents and the flag warning system. *Aquatics International.*

Fletemeyer, J. (2017). Editorial: The impact of beach sand nourishment on public safety. *J. of Coastal Research.*

Fletemeyer, J., Dean, R., and Hudalgo, L. (2012). Physical, demographic, and cultural factors contributing to fatal drownings on beaches located on the Pacific Coast of Costa Rica. *J. of Coastal Science.*

Keshlear, B. (1999). Drowning intervention: An Army Corps of Engineers perspective, in ed. J. Fletemeyer, *Drowning, New Perspectives on Intervention and Prevention*, pp. 165–177. New York: CRC Press.

Leatherman, S., and Fletemeyer, J. (2012). *Rip Currents, Beach Safety and Modeling.* New York: CRC Press.

8 Commercial Whitewater Rafting

Julie Munger and Abigail Polsby

CONTENTS

Introduction 170
   Who's Going Rafting? 172
  Types of Hazards 174
   What Is Whitewater? 174
    Whitewater Class Determination 175
   Whitewater and Hydrologic Features Defined 176
    Laminar Flow 176
    Helical Flow 176
    Eddy and Eddy Lines 176
    Waves 177
    Hydraulic or Holes 177
   River Hazards 177
    Fluctuating Water Levels 177
    High or Low Water 178
    Swimming 178
    Strainer 178
    Sieves and Entrapments 179
   Environmental Hazards 179
    Cold Water 179
    Rigging and Ropes 180
   Previous Medical Conditions 180
Safety Standards 180
  River Guide 181
   River Guide Training 181
  Safety Equipment 183
   Automated External Defibrillator (AED) 183
   Personal Safety Equipment 184
Risk Management 185
  Participant Limits 185
  Flow Limits 186
  Guide Selection 186

Emergency Action Plans, Hazard Mitigation, and Safety Briefing 186
    Emergency Action Plans (EAPs) 186
    Safety Briefing 187
    Boat Order on the River 187
    Inflatable Kayaks as Part of Commercial Rafting Trips 188
Types and Rates of Accidents 189
    Fatalities 189
        Causes of Whitewater Rafting Deaths 189
    Injuries 190
Discussion 191
References 192
Appendix A 192
    Grand Canyon National Park–Colorado River General Trip Gear 193
    Canyonlands National Park Required Equipment for River Trips 193
Appendix B 194
    Suggested First Aid Items 194
    BLM New Mexico Rescue Equipment Requirements 195
Acknowledgement 195

INTRODUCTION

The commercial whitewater rafting and kayaking industry has grown substantially since its humble beginnings in the late 1950s. In 2013, the number of Americans participating in whitewater rafting was an estimated 3.8 million (Outdoor Foundation, 2014).

Some commercial rafting trips are only a couple of hours and others are two to three weeks long in extremely remote areas. While all bodies of water present drowning and injury hazards, whitewater rivers present a particularly unique environment with fluctuating volumes of water, current, complicated hydrological patterns, and obstacles. In addition, each river is unique, shaped by characteristics of the landscape, which presents its own challenges and rewards for whitewater rafters.

Few minimum safety standards are set within the commercial whitewater industry ("the Industry") and the Industry is almost entirely self-regulated. It is perhaps because of the wide variance of characteristics between regions and rivers across the nation that suitable standards have not been settled upon. Professional whitewater guides receive varying levels of training from two general sources: (1) "In-house" training by their commercial employer company and (2) by attending third-party courses taught by swiftwater rescue and wilderness first aid–specific companies.

Though the risks of whitewater may appear high, a well-trained guide can minimize the risk to such a degree that participants often complete a trip with a feeling that it was *too* easy.

One of the great challenges of creating this chapter was the lack of accident data collected or revealed in the Industry. Companies and private governing agencies are extremely protective of accident statistics. In the past, evidence is present that what accident data was available was misused in national publications and misrepresented the risk associated with commercial river trips (Griffen, 2006). One article stated a high rate of fatalities on whitewater trips but did not specify that its data source included both private and commercial trips. Even though the statements made by Griffen were rebutted by another article (Brown), the damage was done and the Industry became very tight-lipped about accidents (Brown, 2007). While some data can be extrapolated from the scant information the Industry reveals, the gap of information is the Achilles heel of the Industry's effort to progress towards a higher safety record.

The purpose of this chapter is to build an understanding of the whitewater industry, its environment, clientele, standard practices, and, when applicable, illustrative case law and outcomes of particular situations. Because of the dynamic nature of water and rivers, each accident case needs to be reviewed independently with special consideration to the unique setting of the particular river; level of guide experience, training, and skill; and the practices of the commercial company, including liability release forms.

This chapter sets out to prepare for such a review by presenting the following basic information:

- Demographics: Who participates in commercial whitewater rafting trips.
- Hazards: Defining whitewater, Industry terms, and the types of hazards rivers present.
- Safety standards: The safety standards to which commercial whitewater rafting trips generally adhere.
- Risk management: Standard practices used to reduce risk in the commercial whitewater industry.
- Types and rates of accidents.
- Overview of current Industry standards and what the Industry could improve upon to reduce accidents and injury.

While the safety of each client is in the hands of his or her guide, ultimately every person who embarks on a commercial river trip is presented with the inherent risks of the activity and must make the conscious choice to accept and assume that risk. Accidents do happen! The river is a dynamic force and there is always a possibility of a tree falling into a channel or the raft bumping a rock, turning a comfortable seat on a raft into an intense submersion into turbulent waters. While other aquatic environments have more obvious hazards and easily mitigated danger, a riverine system has constantly changing dangers that go along with the great rewards that come with experiencing a wild environment.

Industry companies take on the responsibility of providing adequately trained guides, quality boats (rafts), safe rigging, and an adequate emergency action plan. As long as guides act responsibly, are well trained, and do everything they can to prevent an accident from happening, neither they nor the rafting company should have to face the severity of a lawsuit. However, in the case of a death or severe injury, a lawsuit is nearly inevitable.

While whitewater rafting clients deserve every possible safety precaution, they must recognize that the river is a dynamic and unpredictable place and accept the risk that something may go wrong. They acknowledge the dangers and accept the risks involved when they sign a very specific liability release waiver.

In the case of *Espinoza v. Arkansas Valley Adventures, LLC*, Dist. Court, D. Colorado 2014, the court found in favor of Arkansas Valley Adventures when a participant drowned after falling out of a raft and being swept into a submerged tree. The district court concluded that the victim, who had signed a liability release waiver, had assumed the risks inherent with whitewater rafting.

A capsized raft in which all paddlers are successfully rescued quickly can be a highlight of a river trip. Many people are rescued with smiles and laughs. Alternatively, a combination of poor training, lack of pre-planning, or just plain bad luck can turn an otherwise fun swim into a terrifying experience or worse, a death.

Who's Going Rafting?

Whitewater rafting is an experience that facilitates teamwork and having fun in an outdoor environment where clients are removed from the everyday pressures of our technological world. Because of its wide appeal, casual participants from four to ninety years of age participate in whitewater rafting trips each year. Somewhere between 1.3 and 1.6 percent of all Americans over age six went rafting from 2010 to 2012. In 2012, an estimated 3.7 million people went rafting (Outdoor Foundation, 2013). This is a slight drop from 2010, when an estimated 4.5 million people participated in rafting trips. The total number of estimated rafting outings in 2012 is 8.9 million trips (Outdoor Foundation, 2013).

Whitewater rafting demographics tend to follow the U.S. population demographics: 55 percent male and 45 percent female (Fiore, 2003). Thirty-eight percent of rafters went on only one outing per year and 68 percent made three outings or less (Outdoor Foundation, 2013).

Below are basic demographics about the whitewater rafting participants in the United States for 2012 (Outdoor Foundation, 2013).

Age	
8 percent	6–12 years old
13 percent	13–17
13 percent	18–24
37 percent	25–44
28 percent	45+
Ethnicity	
6 percent	African American
6 percent	Asian/Pacific Islander
79 percent	Caucasian
5 percent	Hispanic
5 percent	Other
Income	
15 percent	<$25,000
17 percent	$25,000–$49,000
21 percent	$50,000–$74,999
13 percent	$75,000–$99,999
35 percent	$100,000+

Education

20 percent	1–3 years of high school or less
15 percent	High school graduate
23 percent	1–3 years of college
26 percent	College graduate
14 percent	Post graduate
3 percent	Not specified

Participation in rafting across the country is not uniform. The Outdoor Foundation broke down the United States regionally and represented the percentage of each region's population participating in rafting. The highest concentration of rafting participants is in the Mountain region followed by the East South Central, with the least participation in the West South Central (Outdoor Foundation, 2013)

Pacific (California, Oregon, Washington)
  Participation rate: 1.6 percent
  Percent of U.S. participants: 17.3 percent
Mountain (Nevada, Idaho, Arizona, New Mexico, Utah, Colorado, Wyoming, Montana)
  Participation rate: 2.5 percent
  Percent of U.S. participants: 12.1 percent
West South Central (Texas, Oklahoma, Louisiana, Arkansas)
  Participation rate: 0.9 percent
  Percent of U.S. participants: 7.0 percent
West North Central (North Dakota, South Dakota, Nebraska, Missouri, Iowa, Minnesota)
  Participation Rate: 1.4 percent
  Percent of U.S. participants: 6.3 percent
East North Central (Wisconsin, Illinois, Michigan, Ohio, Indiana)
  Participation rate: 1.3 percent
  Percent of U.S. participants: 12.8 percent
East South Central (Kentucky, Tennessee, Mississippi, Alabama)
  Participation rate: 2.1 percent
  Percent of U.S. participants: 8.5 percent
South Atlantic (Delaware, Maryland, Virginia, West Virginia, North Carolina, South Carolina, Georgia, Florida)
  Participation rate: 1.5 percent
  Percent of U.S. participants: 19.3 percent
Middle Atlantic (New York, Rhode Island, Pennsylvania, New Jersey)
  Participation Rate: 1.4 percent
  Percent of US Participants: 12.9 percent
New England (Maine, New Hampshire, Vermont, Massachusetts, Connecticut)
  Participation rate: 1.2 percent
  Percent of U.S. participants: 3.9 percent

TYPES OF HAZARDS

Whitewater rivers are unlike other aquatic environments. Water parks, pools, lakes, and reservoirs have well-defined and obvious hazards, like diving into a shallow pool. Rivers are more similar to the coastal ocean where water levels are in constant flux and environmental conditions play a major role in identifying potential hazards. Constant hazards exist on rivers, such as cold water or undercut boulders, but variable hazards like weather, dam releases, and flash flooding create the need for a deeper understanding of how changing conditions change the hazards. It is important to understand the basic hydrological principles and terminology of rivers to fully understand their hazards.

River guides must have a thorough understanding of hazards and constantly evaluate and reassess them in changing conditions. A thorough understanding of the hazards leads to a better assessment and employment of appropriate plans to avoid these hazards. There are several ways a river guide can avoid potential hazards: (1) maneuvering around the hazards, (2) keeping clients in the boat and avoiding swims (once a participant is out of the raft and in the water it is referred to as a *swim* and they are referred to as a *swimmer*), (3) portaging around the hazards, or (4) cancelling the trip if water levels present too high of risk to proceed with a trip.

Most hazards are not a problem if the boat is captained by a well-trained and responsible guide, and all passengers are trained and have the ability to stay in the boat.

WHAT IS WHITEWATER?

Whitewater occurs when river water is either falling down a steep enough gradient to create turbulence to a degree that air is entrained into the waterbody or the water is colliding with obstacles, causing it to erupt into a bubbly or aerated and unstable current appearing white.

Rapids are formed by the combination of gradient, obstacles, volume of water, and constriction. As water flows downhill, it gains momentum. When this volume of water goes over obstacles, such as boulders, it creates hydraulic features. Flow in natural rivers is characteristically nonuniform and unsteady.

Whitewater is rated in level of difficulty by being categorized into one of six classes with Class I being the easiest and Class VI being either impossible to pass safely or the most difficult. Classes are applied to individual rapids and sections of river. For example, a section of river that is called Class III may contain mostly Class II whitewater punctuated by two Class III rapids. Class determinations are made by assessing a combination of factors. These factors differ whether rating an entire section of river or an individual rapid. To assess an individual rapid, the following are considered: Number and type of maneuvers that are needed for navigating the rapid and severity of potential consequences if someone falls out of a raft. To rate an entire section of river, the following factors are considered: Rating of the most difficult rapid on the section, average rating of rapids on the section, and distance from additional medical or rescue help.

There is some debate on the finer details of how these classes are determined, but the following descriptions from the International Scale of River Difficulty used in the United States and produced by American Whitewater, a nonprofit organization and the largest public source of information on whitewater statistics, are commonly accepted in the Industry.

Whitewater Class Determination

Class I (Easy): Fast-moving water with riffles and small waves; few obstructions that are all obvious and easily missed with little training. Risk to "swimmers" is slight; self-rescue is easy.

Class II (Novice): Straightforward rapids with wide, clear channels that are evident without scouting. Occasional maneuvering may be required, but rocks and medium-sized waves are easily avoided by trained paddlers. Rapids that are at the upper end of this difficulty range are designated Class II+.

Class III (Intermediate): Rapids with moderate, irregular waves which may be difficult to avoid and which can swamp an open canoe. Complex maneuvers in fast current and good boat control in tight passages or around ledges are often required; large waves or strainers may be present but are easily avoided. Strong eddies and powerful current effects can be found, particularly on large-volume rivers. Scouting is advisable for inexperienced parties. Injuries while swimming are rare; self-rescue is usually easy but group assistance may be required to avoid long swims. Rapids that are at the lower or upper end of this difficulty range are designated Class III- or Class III+, respectively.

Class IV (Advanced): Intense and powerful but predictable rapids requiring precise boat handling in turbulent water. Depending on the character of the river, it may feature large, unavoidable waves and holes or constricted passages demanding fast maneuvers under pressure. A fast, reliable eddy turn may be needed to initiate maneuvers, scout rapids, or rest. Rapids may require "must make" moves above dangerous hazards. Scouting may be necessary the first time down. Risk of injury to swimmers is moderate to high, and water conditions may make self-rescue difficult. Group assistance for rescue is often essential but requires practiced skills. For kayakers, a strong roll is highly recommended. Rapids that are at the lower or upper end of this difficulty range are designated Class IV- or Class IV+, respectively.

Class V (Expert): Extremely long, obstructed, or very violent rapids which expose a paddler to added risk. Drops may contain large, unavoidable waves and holes or steep, congested chutes with complex, demanding routes. Rapids may continue for long distances between pools, demanding a high level of fitness. What eddies exist may be small, turbulent, or difficult to reach. At the high end of the scale, several of these factors may be combined. Scouting is recommended but may be difficult. Swims are dangerous, and rescue is often difficult even for experts. Proper equipment, extensive experience, and practiced self-rescue skills are essential.

Class VI (Extreme and exploratory rapids): Runs of this classification are rarely attempted and often exemplify the extremes of difficulty, unpredictability, and danger. The consequences of errors are severe and rescue may be impossible. For teams of experts only, at favorable water levels, after close personal inspection and taking all precautions. After a Class VI rapid has been run many times, its rating may be changed to an appropriate Class V rating.

WHITEWATER AND HYDROLOGIC FEATURES DEFINED (FIGURE 8.1)

Laminar Flow

Laminar flow describes how water flows in different layers. A whitewater raft sits on top of the water riding the current of one layer. When a person falls out of a raft, they enter a deeper layer of water that is traveling faster than the raft. The current speed that their body is submerged in can carry them away from the boat and their distance from the boat could grow quite quickly.

Helical Flow

Helical flow is caused by water colliding with either itself or the river bank. This creates spiraling currents that present themselves in turbulent boils of water. Helical flow is created when water flows along the bank of the river, and at higher flows can be physically noticed, as it tends to push out toward the middle of the channel. It can also be felt anywhere that the water of the river is flowing in different directions, such as an eddy line. The more helical flow created in the river channel, the harder it can be to seek refuge in an eddy or get to shore.

Eddy and Eddy Lines

As water flows downstream into an object, such as a rock or gravel bar, it piles up on the upstream side creating a slightly elevated water level before flowing around the sides of the object. On the downstream side the water level is lower. Because of the

Whitewater and hydrologic features defined

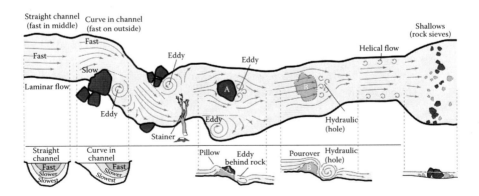

FIGURE 8.1 Hydrologic features seen in whitewater.

Commercial Whitewater Rafting 177

inequality of levels, water from downstream starts to flow upstream to fill in the void behind the rock. This creates either countercurrent or calm water behind the object. Eddies create spots where kayaks, rafts, and canoes can pull into and avoid moving downstream.

An eddy line is the meeting point of currents between an eddy and the flowing part of the river. The river is flowing in one direction and the eddy is flowing in the other direction. Where these opposing currents meet next to each other they form an eddy line, which can create turbulent, swirling water and occasionally require effort and speed to cross out of an eddy into the main current.

Waves
Waves can be formed by water either flowing over obstacles or from the compression of water between the banks of the channel. These features are often used as "play" spots for rafts and kayaks, which can actually surf the wave. It is common for waves to pulse in rivers, meaning they remain in the same place but grow larger and occasionally collapse on themselves.

Hydraulic or Holes
A hydraulic is created by water pouring over an obstacle, such as a rock, that is near the surface so powerfully that as the water pours over the downstream side of the obstacle it creates a void or vacuum to which the water moves upstream into the void, creating a hole. As the water moves below the surface it hits the bottom of the river and can recirculate upward and then back into the mouth of the hole creating a recirculating pattern that can be difficult to escape in a raft or as a swimmer. Irregularities in the obstacle create slight deviations in the recirculating pattern where a swimmer can escape.

The steeper the water falls over the obstacle, the more of a recirculation pattern can be noticed in a hole. Some holes have a big green flow of water on the surface over the rock and out the backside with a big pile of whitewater to break through, which are very fun for whitewater rafters. When rocks have a very steep drop behind them, a recirculating pattern can develop with little current flowing through downstream on the surface. This creates a situation that can stop the raft, turn it sideways, and capsize. The intensity of the hole is dependent upon the flow of the water, and the size and shape of the rock the water flows over.

RIVER HAZARDS

Fluctuating Water Levels
Water levels can change quickly and dramatically in rivers, creating flash flooding within seconds in what was a nearly dry river bed. Many rivers in the United States are controlled by dams that dictate flows on a daily and sometimes hourly basis. Dam-controlled rivers can lead to paddlers either being stranded without enough water to paddle to their final destination or inundated by high water if they are unaware of the dam-release schedule.

With different water levels come different hazards. A river at low flow with shallow water and technical rock-filled rapids presents entirely different hazards than a river with a large volume of water with a wide channel and deep, turbulent rapids. Both of these environments can be exceptionally enjoyable for whitewater enthusiasts.

High or Low Water

Hazards change at high and low water. At low water more rocks and features are uncovered. This usually means that the boat has to do more maneuvering. Rapids tend to change at low water more than they do at high water. Guides need to have skills to navigate their boat safely in both high water and low water.

If a river bed narrows or the volume of water rises, then the water generates faster current and more force. As the volume of the river increases relative to the width of the river banks, the speed and intensity of the hydrology increases. This creates "high water" conditions which require a particular set of skills and precautions.

In a large-volume river such as the Colorado through Grand Canyon, the rapids are relatively short, with long flat water areas immediately downstream. Falling out of a raft in these conditions requires the swimmer to remain calm for a short amount of time until they flush out of the rapid and can be rescued in calmer water.

In other rivers, such as the Yampa River in Utah, rapids are long and continuous at high water flows. In this condition, it is especially important that a very close boat order is maintained, and all precautions are taken to avoid areas that may cause a boat to capsize. Falling out of a raft in high water requires an intense concentration on rescuing the swimmer(s) because large-volume rivers may contain more powerful hydraulics. High-volume water can push a swimmer underwater and result in what is referred to as a "flush drowning" or a drowning caused not by one constant submersion, but by repeated and violent submersions that result in too few opportunities to breathe and eventually totally exhaust the swimmer.

Swimming

Almost all rafting accidents occur when a person falls out of the raft. It should be the guide's priority to get swimmers out of the water. The safest place is always *out* of the water. The swimmer's priority is to get back to the boat and get in, get on top of an upside-down boat, or swim away from hazards. Each participant should have a reasonable idea of how to self-rescue. Though novice rafters may be in too much shock the first time they fall out to quickly assist in their own rescue, they should be instructed on how to actively participate in their rescue if they find themselves in the water.

Strainer

Any object in the current that allows some water to flow through it but restricts passage of a person or boat is considered a strainer. Most strainers are trees that have fallen into the river. The trunk and branches extend above and below water and for the untrained, do not appear to be a hazard. Often people think they can hold onto a branch to stop them, or casually approach the strainer thinking that they can either climb on top of it or flush under it. It takes very little current speed to trap a person under the water in a strainer. Trees may fall into rivers overnight and create sudden hazards where the day before, there was none.

Sieves and Entrapments

Like strainers, sieves and entrapment hazards are caused by obstacles in the water that have the potential to catch hold of or entrap part or all of a person's body or boat and hold it. Sieves are created by obstacles in the river, such as two rocks with a narrow gap between them or a crack in the riverbed. Entrapments occur when a person or part of them becomes entrapped by any obstacle, such as a foot getting stuck under a rock or caught in fishing net, rope, discarded metal, or larger body-entrapping situations such as undercut rock ledges or concrete and riprap road embankments along the sides of the river. These roadside undercuts often occur because the river erodes the soft riverbank below the concrete stabilization slab, potentially leaving behind a cavernous trap depending on water level.

If someone falls out of the raft they should immediately tuck into a ball until coming to the surface to avoid a body part becoming trapped. Once on the surface they should keep their feet on the surface too. This is very difficult for the untrained, as survival instinct tells them to push for the surface. When someone tries to stop themselves in the current by putting their feet down, or push off the bottom to get air, the potential for foot entrapment exists.

Rescues conducted on entrapment victims can be extremely difficult due to the problem of accessing the person mid-stream.

Environmental Hazards

There are environmental and shore-related hazards that come into play more often on multi-day trips. These include heat, sun, dehydration, rattlesnakes, poison oak or ivy, uneven ground, bee stings, and so on.

Cold Water

The effects of cold water submersion on the human body are extensive (Giesbrecht, 2006). In cold water the airway constricts, the vessels in the circulatory system constrict, and rapid cooling of the extremities can lead to a decreased ability for self-rescue. The effect of cold water on the person entering it will be highly dependent on the safety equipment provided. Thermal protection from a wetsuit or drysuit will decrease the long-term impact of the water. Only a drysuit can insulate a person from the shock of the rapid change in temperature. In Alaska, where water temperatures can be in the low 40s, outfitters routinely supply clients with drysuits whenever there is a rapid large enough to risk being thrown from the boat. Thermal protection from wetsuits is often provided when the water temperatures warrant it. A wetsuit will not protect someone from the initial shock of the cold water, but will help keep someone's body temperature up as they recover from the swim.

Cold water can cause both cold water shock and hypothermia if the passenger is not rescued and out of the water as soon as possible. Reaching the point of hypothermia usually requires an extended swim, and is more likely to occur after the person is out of the water (Giesbrecht, 2006).

Rigging and Ropes

One of the biggest hazards on the river is the actual rigging in the boat itself, and the stowage of the safety gear. Drownings and near-drownings have been caused by entrapment in a rope or a boat's rigging (Lower Gauley, 2008). Rigging needs to be secure, and safety and bowlines need to be stowed in a way that they cannot self-deploy under any circumstances.

PREVIOUS MEDICAL CONDITIONS

With the age of participants increasing in the industry, the number of incidents caused by existing medical conditions is also rising (Brown, 2015). The shock of falling out of a raft into cold water with the added exertion of swimming through whitewater can induce existing medical conditions such as cardiac arrest, asthma, or seizure.

SAFETY STANDARDS

In the United States, training levels for safety, navigation, and rescue are largely Industry-driven and are not uniform. Each individual outfitter must meet some minimum safety standards required by the agencies that govern the land that the rivers run through. The safety requirements vary, and there is not a particular safety standard that is followed. The safety standards that are set by states and agencies are often the absolute minimum. Some countries, such as New Zealand, have strict guide licensing procedures that individuals must follow before they can take passengers downstream. New Zealand river guides must perform skills tests and demonstrate navigational, leadership, first aid, and rescue skills. This is not the case in the United States. While a basic level of first aid is required nationally, other skills like rescue, navigation, and leadership are left largely to the outfitter's discretion.

In the United States, different regions or even sections of rivers have different requirements for commercial guides. The National Park Service has guide requirements on all the sections of rivers that run through the lands they manage; however, each park has different requirements. For example, on a multiday Grand Canyon trip, a guide must have a minimum of forty hours of wilderness first aid training, CPR, and a certain number of trips down the river, but rescue training is not required. Where the Snake River runs through Teton National Park, each guide must have a minimum of sixteen hours of Wilderness First Aid, at least one guide on the trip must have had a river or swiftwater rescue training sometime in their lifetime, and each guide must have seen the river a minimum of three times.

Other land management groups and jurisdictions, such as Bureau of Land Management or the U.S. Forest Service, have different commercial river guide requirements for different sections of rivers on the lands they oversee. For example, California State Parks require Basic First Aid, CPR, and swiftwater rescue training on the rivers it manages, but it only requires copies of certifications for first aid and CPR. El Dorado County in California, home to one of the West's most popular day runs, the South Fork American River, requires that commercial outfitters have at least one guide per trip who is trained in swiftwater rescue.

Safety standards for commercial rafting include safety gear, personal protective equipment to be worn by passengers, and guide training requirements. Usually, safety gear required to be carried on each trip is very straightforward and basic (throw ropes, extra paddles, extra life vest, and first aid kit). Guide training requirements generally mean a guide has made a specified number of trips down the river and occasionally receives additional training in swiftwater rescue or wilderness first aid to be the trip leader.

River Guide

The river guide is the person who is physically in charge of a whitewater raft. The raft is usually propelled either by individual participants in the boat paddling under the command of the river guide or a raft in which the river guide rows the boat with long oars. Occasionally, these two techniques are combined into what is referred to as an oar/paddle combination. In a paddle boat, the river guide is responsible for directing passengers to paddle the boat by shouting commands from the back of the boat. In an oar boat, the passengers are instructed on how to hold on and stay low in the boat while the river guide oars the boat down the river. In both situations, the river guide is also responsible for safely rigging the boat, stocking rescue and first aid gear, stowage of any passenger personal items, and disseminating additional safety information.

It is also the job of the river guide to give a safety talk to all participants on the rafting trip before embarking on the trip. The river guide assures that this safety information is understood and adhered to during their time with the passengers in their raft.

River Guide Training

The first step in a river guide's training is usually a guide school. These trainings are offered in-house by commercial outfitters as a way to train new guides and are often a type of job interview for new guides. These guide schools are often, but not always, offered before the regular rafting season. Commercial outfitters often use returning guides as guide school instructors. Most river guides are hired seasonally.

During guide school, new candidates are taught the basics of hydrology, maneuvering the boat in current, paddle captain responsibilities, basic rowing, safety procedures, cooking procedures, equipment maintenance and care, basic rescue procedures, and other jobs relating to the day-to-day operation of a river trip. These trainings usually last from five to ten days. Often, new recruits are selected based on their work ethic and their aptitude for basic river navigation and communication and teamwork with others.

Individuals selected from the guide school usually begin employment with an internship of varied length depending on skill level. During this training, the guide "shadows" other guides by being a paddler in their boat; taking turns at captaining the boat; and continuing to learn about safety, navigation, and company policies.

Ultimately, the outfitter and lead guides determine when the new candidate is ready to captain her or his own boat. As noted above, some government agencies in the United States set a minimum number of trips down the section of river to be guided that a new guide must complete.

New Zealand has strictly objective assessment criteria that must be passed before any guide can operate as a guide. The assessment includes all the previously noted navigational skills plus rescue and first aid skills. The International Rafting Federation (IRF) has established a system that accredits raft guides for varying classes of rivers as well as accrediting trip leaders, assessors, and instructors. While the IRF has a very strong following in other countries, it is still struggling to break into the U.S. market. Currently, assessment of skills related to guiding boats down the river is left entirely to the hiring outfitter. The outfitter also determines whether other training, such as wilderness first aid or river rescue, is required.

Accidents can happen in any river, and drowning happens within minutes, so there is no reason that every commercial guide should not be trained to handle medical and rescue emergencies.

Encouraging guides to become comfortable on different types of rivers—small volume, large volume, glacial run off, desert runs, and wooded sections—will help them be a better guide. This helps them to be more flexible and responsive when something changes. Some companies encourage guides to see different types of rivers and even help aid them with opportunities to experience different rivers. Companies understand the importance and value of a well-rounded guide, not just someone who knows one section of one river really well.

River or Swiftwater Rescue Certification

There are organizations that teach river rescue that is specific to river guides in the United States. These include the American Canoe Association (ACA), Sierra Rescue, Rescue 3, and others. Some offer internationally recognized certification, while organizations such as the ACA offer workshops. As with any industry, some organizations have much higher qualifications and standards.

CPR

Most companies require this from their guides. Some companies accept online certifications whereas others require a course with an instructor. It is our strong opinion that online courses should not be an acceptable way to practice and become certified in CPR.

First Aid, Wilderness First Aid, Wilderness Advanced
First Aid, Wilderness First Responder

Standard first aid courses are often not specific enough to the environments that river guides find themselves in. Standard first aid courses teach students to call 911 and assume that help is close by, which is rarely an option on a river trip. Wilderness medicine courses are based on being one hour or more away from definitive medical care. While a wilderness-specific first aid course is not required of many river guides, it would behoove any outfitter who operates in remote locations to begin requiring it. While a river guide cannot be expected to be a medical professional, they should possess the skills to understand a good course of action for treatment in the river environment. The more remote and the longer the river trip, the more first aid training may be necessary.

Safety Equipment

All commercial river trips carry safety equipment. There are even some areas where river runners are not only required to carry, but are physically checked to confirm they are carrying particular equipment before they launch on their adventure. A list of safety gear required by Grand Canyon National Park Service is included in Appendix A. It should be noted that the safety standards set by the Grand Canyon outfitters are some of the highest because of the combination of its remote nature and difficulty of whitewater. These outfitters are the exception rather than the rule.

When it comes to first aid kits, again, there is no standard. Some river companies have large kits that are very well marked, easy to get to, and stocked with everything you can need. Others supply small, hard-to-find kits with insufficient supplies for anything major involving more than one or two persons. There are a few first aid kit companies that market specifically to river trips. They supply well-thought-out waterproof kits that are created for a variety of needs (i.e., eight people for seven days).

There are certain supplies that could be added into the kit depending on the level of first aid training the river guides or runners have. For example, wilderness first aid– and first responder–trained individuals are taught the use and importance of epinephrine and other potentially life-saving drugs. These do not come stocked in pre-made kits and need to be added in. A list of suggested first aid items for river trips can be found in Appendix B.

Automated External Defibrillator (AED)

Automated external defibrillators, or AEDs, are not on any recommended first aid supply lists. A google search for "river guides and AED" will reveal that many companies train their guides in CPR/AED but do not actually bring AEDs on their trips. It is both financially impossible and unrealistic for a large outfitter to carry an AED on every river trip.

AEDs have been successful in resuscitating at least three teenage passengers on river trips. There are no reported cases of adult resuscitation with an AED on a river trip. In adults, the problems or conditions that cause heart attacks remain after potentially being revived with an AED and they will not fully recover without supportive therapy such as IV cardiac medications. Most river trips take place in what is considered "wilderness" for first aid care, meaning that they will not get to definitive care in less than one hour. The general thought in the wilderness medical community is that without the IV medications, AEDs alone don't save people.

A recommendation could be made for stashing AEDs along the popular commercially run rivers as some search and rescue teams do with transportation litters and first aid kits. Examples of these caches can be found on the South Fork American and Tuolumne Rivers in California.

Participants with any life-threatening conditions should always inform their guides before agreeing to a river trip. In fact, most registration questionnaires for rafting trips request this information and it is often stated in the release form that participants are required to divulge pertinent health information and medications. A rafting passenger with a heart condition assumes a greater risk when they decide to participate in more difficult or remote trips. Some groups and clients accept this risk and have been known to come on the trip with their own AED.

Personal Safety Equipment

Aside from group gear that is carried and provided for the trip participants, it is not required by all companies but recommended that each guide should have on their person a minimum amount of gear to perform basic rescue procedures should they become separated from their raft. During swiftwater and river rescue courses, guides are taught and encouraged to have the following on their person, or in their boat: a throwbag, plus the 4-3-2-1 system.

4-3-2-1 safety gear for rafters:

- 4 carabiners
- 3 pulleys
- 2 prusik loops
- 1 piece of webbing (12–20 feet)

Most commercial river companies supply the following river rescue equipment:

- 1 throwbag per boat
- 1 × 200–300 feet static rope
- 3–4 carabiners
- 2–3 pulleys
- 2–4 prusik loops
- 2–4 1-inch piece of webbing ranging from 6–12 feet

Commercial companies would benefit from following the advice of some private boating clubs. Below is an excerpt from the group preparedness and responsibility section of the Safety Code for American Whitewater created by Charlie Walbridge and Mark Singleton in 2005.

Group equipment should be suited to the difficulty of the river. The group should always have a throw-line available, and one line per boat is recommended on difficult runs. The list may include carabiners, prussic loops, first aid kit, flashlight, folding saw, fire starter, guidebooks, maps, food, extra clothing, and any other rescue or survival items suggested by conditions. Each item is not required on every run, and this list is not meant to be a substitute for good judgment.

New equipment is being designed all the time and as these get released, these lists will change. New technology has come out even in the last few years making people feel safer and more connected such as the Spot Device or DeLorme's InReach for communication. Commercial trips should always carry a means of communication to the outside world and know the appropriate times to use it.

RISK MANAGEMENT

Rivers are constantly changing and there are so many variables that go into a river trip. People can attempt to plan for all the known possible situations that may happen on a river trip, but sometimes crazy, unpredictable stuff happens.

How can companies work on improving their risk management? Currently companies have the following strategies to deal with risk management:

Participant Limits

Outfitters make trip recommendations to the public based on the physical abilities and medical conditions of the participants. While physical abilities must be assessed at the river site, medical conditions and important factors, such as the weight and medical history of the individual, can be assessed prior to departure. While rafting companies often specify minimum age requirements, maximum age limit requirements are not mandated due to age discrimination laws. All passengers must, however, properly be fit in a PFD (personal floatation device or life vest). This usually creates an upper weight limit of around 250 pounds; however, distribution of that weight can vary this upper limit.

Denying a person a spot on a trip based on the belief that they are not fit for a trip is not without repercussions. Respected outfitters have received threats from fit passengers who may be in their 80s but want to participate in such remote places as the Grand Canyon on multiday or multiweek trips. If these people are cleared by their doctors to participate, then it is relatively impossible for an outfitter to deny them access to the experience without risking an age discrimination lawsuit.

For smaller children, acceptance on a trip should be made based more on the size of the child and availability of a securely fitting PFD than the age of the child—not unlike car seats being based on weight and not age. Most outfitters across the United States recommend a minimum age of four to five for Class II, seven to eight for Class III, twelve to fifteen for Class IV, and sixteen to eighteen for Class V rapids. This seems to be an average based on a random sample of outfitters conducted by the authors of this chapter.

General limitations for people participating on rafting trips are the ability of the person to fit well in a PFD, the ability of the participant to self-rescue, the ability of the participant to float safely while being rescued by someone within the river party, and survivability in the temperature of water in which they may find themselves. Ability to self-rescue becomes more important as the difficulty of the river increases. As the difficulty of rapids increases, the importance of actions by the participant becomes more important. For example, an overweight person with cardiac issues must be able to relax in a stressful environment. A person must have enough self-control to keep their limbs away from the bottom of the river and follow instructions.

Flow Limits

In the United States, the volume of water in rivers is measured in cubic feet per second (the number of cubic feet of water moving past a fixed point in one second). It may also be measured on fixed gauges in height of feet. Most commercial companies have upper cut-off volume or "flow" limits. This means that the company won't run a section of river if the flow reaches their designated high point due to safety issues. They also have lower cut-off limits to avoid hazards that might appear when water is too low. See the hazards section for specific notes on high and low water.

Guide Selection

Hiring guides with the appropriate training is the Industry's number-one defense against accidents and injuries. Guides need to be trained on each river section they are going to be guiding. Some companies require a guide to have run a section of river a certain number of times before they are allowed to take passengers. There is usually a final check-off where the guide does the run with a manager or lead guide in their boat before they are allowed to take passengers.

Companies should know that their guides can handle a situation. Training in River Rescue Certification Courses, Wilderness First Aid and CPR courses, Food Handlers Certification, and having natural history knowledge all help guides deal with the emergencies that arise on a river trip and provide a superior rafting experience for clients. River rescue and wilderness first aid courses specifically conduct different types of river emergencies, giving guides essential practice and an opportunity to demonstrate skills to their employer. These help prepare them if something does indeed go awry on one of their trips. Some companies take it a step further and have scenario days where their guide teams or crews work together in rescue scenarios and simulations. This is either done in-house or an outside company is brought in to facilitate this. Other companies have competitions or practice days where they help keep their guides skills sharp. These sessions are usually run by the lead guides, managers, or owners of the river company.

Emergency Action Plans, Hazard Mitigation, and Safety Briefing

Emergency Action Plans (EAPs)

An Emergency Action Plan is an evolving, all-inclusive process of written documentation, verbal communication, discussion, and procedures. The EAP starts with the guide(s) giving a safety talk to all passengers before they embark on a trip. Another guide may even have a checklist to be sure the speaking guide hits all the points. The EAP includes discussion between the guides and their superiors of particular hazards and the plan for dealing with a mishap due to these hazards. Discussion starts at the beginning of the trip and it continues all through the trip as guides check in with each other at major rapids. Part of the EAP protocols might be written out and contained in the first aid kit as a protocol for major medical or rescue emergencies.

The EAP must be on both a microscale and a macroscale. A worst-case scenario must be thoroughly thought out, and resources at each stage of evacuation must be identified. This starts from the most basic process of how to get out of the water to how and where the evacuation will happen, who to call, and knowing what resources are and are not available.

A good EAP discussion or documentation includes descriptions of what to do in situations where there are swimmers, the craft is upside-down, or an individual is separated from the craft. The EAP includes egress and access sites of that particular river, what form of communication to use and who to call in an emergency, and what to do for each particular situation.

Every guide should be familiar with each rapid, where the hazards are, and what the EAP is for each rapid. The guide must also understand the bigger picture of evacuation to medical care such as: Where can a helicopter land? and What resources may or may not be available to help?

EAPs are reviewed and updated through incident reports or safety briefing notes. EAPs increase accident prevention, as outfitters, managers, and river guides are guided through the process of thinking through all the "what-if" scenarios for that particular trip. Every trip is different and that should be acknowledged.

Safety Briefing

These are also referred to as *safety talks*. Every company conducts their own talks with their clients before they get into the boat or leave the shore to enter the river. Because some river trips have many components to their safety briefings, they may split it up into multiple talks. For example, the initial safety briefing will cover how to be safe on the boat and in the river, the next will cover camp safety, and another one before a hike. Additional safety talks may be held before individual rapids while scouting, and occasionally clients have the option to run the rapid or walk around it.

Common components for a water safety briefing are

- PFD—fit, when to wear it, and where to put it when you take it off
- How to sit in the boat
- How to hold on in the boat
- Boat and paddling commands
- Hand signals
- What to do if a client falls out of the boat
- What a throwbag or throw rope is used for
- How to grab a throw rope if one is thrown
- What to do if you get separated from the group

The safety talk is also the time a guide might ask that anyone with health conditions report any medical issues to the guide regardless of whether the liability release waiver does or does not specifically ask about them.

BOAT ORDER ON THE RIVER

The boat order and where the boats are in relation to each other is one of the key safety factors of a river trip. At higher water levels, boats must run incredibly close together to be able to access a swimmer in the water. At low water levels, caution must be taken to avoid running into each other or running over a swimmer. Boats within a river trip must stay in close proximity to each other in order to be effective in any kind of simple or complex rescue or medical problem. It is essential that guide skills and passenger fitness and survivability be taken into account when choosing a safe boat order.

A river guide is always responsible for ensuring the safety of the boat behind him or her. If there is a mishap on one raft, the raft ahead of it alerts the boats ahead by a whistle signal. The downstream raft is often the rescue boat for any swimmers that fall out of the raft behind it. In general, a swimmer floats faster than a raft, although this is entirely dependent on the hydrology of the river.

Either the trip leader or one of the most experienced guides will take the lead position while floating downriver. The first boat and its passengers are the most vulnerable to a flush drowning in high water if the boat flips or has an incident because there is no one downstream to rescue swimmers. Because of this, the second boat in line must also have an experienced raft guide. If there is a problem with the first boat, the second boat must be able to quickly catch up to the first boat. Participants are also selected for particular rafts based on their perceived abilities; stronger paddlers will be in the lead rafts. At high water, lead raft participants must understand the importance of self-rescue. It is similar to being placed in the exit door on an airplane. There are certain tasks the participants need to be able to perform. Fitness, age, and river experience are a few of the things that should be considered.

The last or "sweep" boat usually carries the major first aid kit and additional rescue and communication equipment. The sweep guide has the responsibility of rescuing anyone who may have fallen out of the boat and been left behind. By staying behind everyone else, there is never a chance that the needed equipment has already gone downstream of an emergency. This is critical as the more boats there are on a trip, the more inherently spread out the trip is.

INFLATABLE KAYAKS AS PART OF COMMERCIAL RAFTING TRIPS

Some rafting companies provide inflatable kayaks as part of a commercial rafting trip. This allows participants to have a more unique, challenging, and independent experience. This becomes an additional supervisory challenge for the river guides. In order for someone to participate safely in an inflatable kayak, they must understand how to maneuver the kayak around hazards in the river and self-rescue back into the kayak. This takes focused training by the river guides and is often difficult to do in addition to the other responsibilities of a river guide. The guide will be directing their own raft and paddlers in addition to coaching the paddler in an inflatable kayak. It's typical for inflatable kayakers to follow a specific raft through the rapids. Extra caution must be taken to train the participants to be able to avoid hazards and fit into a safe pattern in the boat order. The number of accidents related to this activity is not included in the scope of this chapter, however, American Whitewater has reported a handful of fatalities due to inexperienced paddlers renting inflatable kayaks without mandatory training or dedicated supervision of the paddler.

TYPES AND RATES OF ACCIDENTS

FATALITIES

Though whitewater rafting has many apparent risks, the chances of dying during a day of rafting are still very low. The actual risk level is difficult to determine with accuracy, however, because reports on fatalities are only submitted voluntarily to American Whitewater or gathered from newspapers or social media. While some local government agencies require accident reports, there is no national repository for them. Because there is no compulsory accident, injury, and fatality reporting requirements, companies in the Industry rarely divulge their own statistics. The information gathered by American Whitewater includes commercial and private rafting trips. A few states or regions have periodically conducted surveys on their own whitewater industries, but no nationwide information is publicly available.

Data from West Virginia from 1984 to 1999 shows there were only 11 fatalities in approximately 2 million commercial rafters, or 0.000005 percent (Mason, 1998).

According to David Brown, President of America Outdoors, the number of fatalities on commercial guided whitewater trips in the United States has ranged between 6 and 10 per year for an estimated 2.5 million user days. The incidence of fatalities is estimated at 1 death per 312,500 user days, or less than 0.00001 percent (Brown, 2006).

Causes of Whitewater Rafting Deaths

The following statistics were assimilated from information gathered by American Whitewater between 2000 and 2014 for commercial rafting trips. Of 125 deaths, 97 (or about 78 percent) occurred while the participant was on a raft trip while the other 22 percent of deaths occurred in other vessels, including inflatable kayaks, canoes, drift boats, and inner tubes (in these cases a guide may or may not have been present, but the equipment came from a rental outfitter). Because these deaths are reported voluntarily to American Whitewater, these numbers do not encompass all whitewater deaths in the United States.

The following list represents the primary causes of death after the individual fell out of the raft, based on the best data available. The list is separated into two categories: Non-entrapment fatalities and entrapment fatalities.

Non-Entrapment Fatalities

37.6 percent—Long swim (This category could contain "flush drowning")
5.6 percent—Heart attack
5.6 percent—No PFD
4.8 percent—Unknown (many of these are described as a combination of causes)
3.2 percent—Hydraulic
2.4 percent—Low-head dam
1.6 percent—PFD fell off
1.6 percent—Head injury
0.8 percent—Other accident
Total for non-entrapment fatalities: 63.2 percent

Entrapments Fatalities

9.6 percent—Swim into undercut
6.4 percent—Foot entrapment
5.6 percent—Entrapment
4.8 percent—Swim into tree
2.4 percent—Tree pin
2.4 percent—Equipment entrapment
1.6 percent—Swim to entrapment
1.6 percent—Pin undercut trap
0.8 percent—Rock sieve
0.8 percent—Swim under raft
0.8 percent—Suction pin
Total for all types of entrapment: 36.8 percent

In summary, the two leading causes of fatalities on commercial rafting trips are long swims and entrapments. Long swims may be complicated with other factors, such as a hydraulics holding a swimmer down longer than they can sustain in combination with a swim. Entrapments are broken down into more specific categories and these also may involve a swim and other factors. While the occasional case exists where someone jumps off a raft, submerges, and simply does not resurface, the causes leading to a fatality usually involve a series of events.

One category missing from these data is whether the swim induced a medical condition like a heart attack. According to American Outdoors President David Brower, heart attacks are becoming more common as a cause of death among commercial rafting participants as the Teneral population ages and as older people are participating in such activities (Brown, 2014). In many of the unknown causes of death category, the victim was rescued and alive when pulled out of the water and died of unidentified or unconfirmed causes on dry land. Some of these causes may have been from a swim-induced medical condition such as heart attack, asthma, or seizure. In future reporting of accidental deaths on whitewater, the category of "swim-induced medical condition" should be included.

INJURIES

While the information available for commercial rafting fatalities is slim, the information on rafting injuries is nearly nonexistent on a national level. The state of West Virginia passed legislation requiring the Industry to report accidents and injuries meeting particular criteria. Without this requirement, there could be no data at all on commercial whitewater injuries.

David Fiore, MD, found that in 2002 there were three to six rafting injuries for every 100,000 boat days. Of those injuries, the number-one injury was contact with a paddle. The second most common injury was received when the rafter made contact with a rock, usually while swimming, meaning they were thrown or tipped out of the boat. These numbers were compiled during the 2002 rafting season.

A 2013 study focused on the popular rafting destinations of the New and Gauley Rivers of West Virginia describes the rates, patterns, and likelihood of injuries based from information gathered on standardized injury report forms for years 2005–2010. During that period, over one million clients participated in commercial rafting trips on these two rivers.

During this six-year period, 205 guests were injured, leading to an injury rate of 20 injuries per 100,000 commercial rafters (or .0002 percent). There were seven fatalities recorded for this same period (fatality rate, 0.68 fatalities per 100,000 commercial rafters).

They found that overall, the number of musculoskeletal injuries comprised the majority of non-fatal injuries. The breakdown of injuries was sprains/strains, 21.1 percent; dislocations, 13.8 percent; fractures, 12.2 percent—totaling 47.4 percent of the non-fatal incidents. Soft tissue injuries comprised 42.3 percent, with 29 percent of these being lacerations and 13.3 percent abrasions.

Almost half (44.3 percent) of these injuries were to the head, neck, and shoulders followed by the lower extremities (foot/ankle/leg/knee/hip 33.9 percent) and upper extremities (hand/wrist/arm 14.3 percent). Their study showed a higher likelihood of injury occurring inside the raft than when a passenger is ejected into the river.

Acute rafting injuries are most often due to contact with another rafter's paddle or other equipment; the next most common injury is the rafter hitting an object while swimming. Chronic injuries are very uncommon in rafting.

The most likely injuries to occur to passengers are a laceration (probability of 0.51), sprain ($P = 0.20$), and dislocation ($P = 0.20$). The main probabilities of the risks of injuries are to the head ($P = 0.24$), shoulder ($P = 0.16$), ankle ($P = 0.11$), and face ($P = 0.13$) of the passenger.

DISCUSSION

Millions of Americans of all ages have the pleasure of participating in whitewater rafting every year and statistics show there are very few deaths or injuries. Rafting with a professional outfitter provides an exceptional opportunity for families and individuals to enjoy time together in a natural environment, which can be both challenging and rewarding. Due to the self-regulation of the Industry, and the lack of oversight on the quality and training of guides, the public has the added responsibility of seeking out safety information about a particular river or outfitter on their own. With very little data available, and a very protective Industry, this information can be challenging to acquire.

The hazards involved in whitewater rafting can be minimized by proactive steps taken by outfitters, including having well-trained guides, water level cut-off limits, safety and prevention practices, and constant vigilance. The public has the responsibility of determining which outfitters have the best practices, and this information is not readily available, as no safety records are published. River trips are advertised by the amount of excitement and whitewater that may be encountered, not the ability of the guides to avoid hazards and keep passengers inside the raft.

The Industry is extremely protective of accident statistics, and the only information available is voluntarily submitted. Statistics about accidents resulting in litigation settled outside of court are not publicly available.

The safety of passengers on a whitewater rafting trip depends on an Industry-driven culture of safety, which includes everything from a good EAP to navigation skills to guides keeping passengers in the boat. When a mishap occurs, safe and quick rescue is often determined by the skill of the guides and the proximity of other boats in the party to the incident. Once participants sign the liability release waiver and push off from shore, they assume the risk that unexpected accidents may happen even though the outfitter and guide have done everything to promote their safety. Participants assume that the beauty and fun of the river trip will captivate them and the guide will be doing everything they can to prevent a fall into the river, an environment that the passenger is minimally trained to survive.

REFERENCES

Brown, D. 2006. Guided Rafting Accident Statistics. http://www.americanwhitewater.org/content/Article/view/articleid/29824.

Brown, D. Phone conversation between David Brown and Haven Livingston on December 16, 2014.

Espinoza v. Arkansas Valley Adventures, LLC, Dist. Court, D. Colorado 2014. http://law.justia.com/cases/federal/districtcourts/colorado/codce/1:2013cv01421/141060/29.

Fiore, D. C. 2003 (December). Injuries associated with whitewater rafting and kayaking. *Wilderness & Environmental Medicine*, 14:4, 255–260.

Foot entrapment by old rope example. http://triblive.com/news/westmoreland/4331550-74/vega-black-rescue#axzz3PgKOBULl.

Giesbrecht, G. G., and Wildernson, J. A. 2006. *Hypothermia Frostbite and Other Cold Injuries: Prevention, Survival, Rescue and Treatment*, 2nd ed. The Mountaineers.

Griffin, D., and Polk, J. 2006 (September 6). Whitewater deaths surge in U.S. CNN. http://www.cnn.com/2006/US/09/05/whitewater.laws/

Lower Gauley River carnage, foot entrapment by raft rigging. 2008 (September 7). https://www.youtube.com/watch?v=WF341zI4eUk

Mason, M. 1998 (August 2). West Virginia whitewater stays the course for 30 years. *Charleston Gazette*, 1c, 5c.

National Highway Traffic Safety Administration. 2012. Fatality Analysis Reporting System Encyclopedia.

The Outdoor Foundation. 2013. A Special Report on Paddlesports.

The Outdoor Foundation 2014. Outdoor Participation Report.

Sayour, G. Whitewater Rafting Death Statistics. http://paddling.about.com/b/2011/06/25/whitewater-rafting-death-statistics.htm.

APPENDIX A

Example of standardized required safety and first aid equipment.

Grand Canyon National Park–Colorado River General Trip Gear

Here is a list that Grand Canyon National Park has listed in their 2014 Commerical Outfitter Regulations (COR).

Raft/Kayak Gear

First aid kit (major)
First aid kit (1 minor kit per additional boat)
Signal mirror (Air Force type)
Signal panels: 2 at 3′ × 8′
Satellite phone/ground-to-air transceiver
River map or river guide books
Watercraft sufficient for size of group
Night navigation lights (if any traveling by night)
Front deck reinforcing (if canoes and kayaks)
4-stroke motors (if motorized trip)
Personal Flotation Devices (PFDs) of Type I, III, or V: 1 per person + 1 extra per 10 people
Throw cushion (Type IV): 1 per boat 16′ or longer
Air pump (1 per trip if inflatable rafts or pontoons)
Boat patching and repair kit
Extra paddles (2 per paddle powered boat)
Fire extinguishers (if motorized trip)
Spare kayak paddle (1 per 4 kayaks)
Spare motor parts (if motorized trip)
Spare oars (2 per oar boat)

Canyonlands National Park Required Equipment for River Trips

Federal and state regulations outline an assortment of equipment necessary for river trips. The following items are required:

- One approved, serviceable Type I, III, or V personal flotation device (PFD): Two for each trip participant.
- One spare PFD: Two for every five people on the trip, or one per boat; whichever is fewer.
- A readily accessible spare means of propulsion capable of maneuvering the vessel (oar, paddle, motor, etc.) for each boat. Low capacity boats designed to carry two or fewer occupants (canoes and kayaks) may carry one spare paddle for every three boats. Commercially made hand paddles are approved for hard-hulled, whitewater kayaks.
- A serviceable Type IV throwable device (throw cushion) for every boat 16 feet or more in length. A commercially made throw bag with at least 40 feet of line is allowed in lieu of a Type IV throwable device.

- Repair kit or kits adequate for repairing the number and types of boats on the trip. Hard-hulled boats may carry epoxy and duct tape or an equivalent means of repair.
- If boats with inflatable components are used on the trip, an air pump or pumps.
- A bailing device or bilge pump for boats that are not self-bailing.
- A first aid kit adequate for the number of trip participants and length of trip.

APPENDIX B

SUGGESTED FIRST AID ITEMS

Items should be neatly stored in an easy to locate and identifiable waterproof container. A first aid kit inventory list should be taped to the inside lid of the container with a Red Cross First Aid Manual or equivalent. The National Park Service highly recommends including all the following first aid items:

Instruments	Description	Uses
Antibacterial soap (Phisoderm, tincture of zephesis, Hibiclens)	8 to 12 ounces	Antiseptic for wounds
Moleskin	1 package	For blisters
Betadine	1 bottle	For cleaning wounds
Band-Aids	36 (1-inch)	For lacerations
Antibacterial ointment	2 tubes	For lacerations and wounds
Butterfly Band-Aids (or know how to make)	18 (various sizes)	For closing lacerations
Carlisle trauma dressing or substitute (feminine napkin, etc.)	3 (4-inch)	For large bleeding wounds
Elastic bandage	2 (3-inch)	For sprains and securing rigid splints
Steri-Pad gauze pads	18 (4-inch by 4-inch)	For large wounds
Steri-Pad gauze pads	18 (2-inch by 2-inch)	For small wounds
Tape, waterproof adhesive	2 (2-inch rolls)	For sprains, securing dressings, etc.
Triangular bandage or muslin pieces	4 (40-inch)	For securing rigid splints, slinging and securing extremities, and protecting dressings from contamination
Roller gauze	5 rolls (2 inch by 5 yards)	For holding gauze pads in place, securing splints and improvising slings
Rigid splint, arm board, Sam Splint	1	For in-line fracture, pressure bandage
Pain reliever (aspirin or substitute)	36 tablets (5 grain)	Headaches, minor pain and fever
Ibuprofen	200 mg tablets	Muscle strains, minor pain, or cramps
Antacid	18 tablets	Upset stomach
Antihistamine	18 tablets	Insect bites, colds, hives or rashes

(Continued)

Instruments	Description	Uses
"Gookinaid" or similar electrolyte replacement drink	1 tub minimum	Relieve or prevent muscle cramps and symptoms of heat exhaustion
Oil of clove	1 small bottle	Relief of toothache
Calamine lotion or cortisone cream	1 bottle	Relief of itching from poison ivy, rash or allergy
Solarcaine	1 bottle	Relief of sunburn pain
Zinc oxide/PABA or sunblock	1 bottle	Prevent sunburn
Benadryl syrup	1 bottle	Minor allergic reactions
Mineral oil or other mild laxative	Small bottle	Constipation
Syrup of Ipecac	Small bottle	Induce vomiting
Kaopectate	Small bottle	Treat diarrhea
Ophthalmic wash or eye drops	Small bottle	Eye wash/irritation
Ear drops	Small bottle	Clogged/infected ears
Water purification tablets	Small bottle	Purify water on side canyon hikes
Eye pad	2	Injured eye
Tincture benzoin	2 small bottles	To hold tape in place and protect skin
Insect repellent	Large can or bottle	Flies, ants, mosquitoes

Note that the above first aid kit list is for a multiday trip and that people and companies doing day trips don't need to carry such an extensive kit.

BLM New Mexico Rescue Equipment Requirements

Required river recreation items (see http://www.blm.gov/nm/st/en/prog/recreation/recreation_activities/boating_and_kayaking.html):

- A first aid kit adequate to handle common river injuries.
- An approved U.S. Coast Guard life preserver, Types I, III, or V, which must be worn by every person in the party while on the river.
- For non–self-bailing rafts, appropriate buckets must be on board each craft.
- A length of rope at least equal to the length of the boat.
- A throw line at least 50 feet in length must be on board each primary boat.
- Patching and repair equipment, including an air pump.

ACKNOWLEDGMENT

Haven Livingston contributed to the research and editing of this chapter.

9 *Breaux v. City of Miami Beach*, 899 So.2d 1059 (Fla. 2005)

Howard Pomerantz

On February 20, 1997, Zachary Breaux appeared to have everything he had dreamed of and worked for within his grasp. The thirty-six-year-old jazz guitarist/composer was becoming established as a star in the jazz/hip hop industry. He had a long-term contract with Zebra Records, a California record company, to record six albums. His recently released third CD, *Uptown Groove*, had just reached number one on the contemporary jazz radio charts, and Zachary was set to begin a worldwide tour to promote the album.

Zachary and his wife, Frederica, a public school teacher, lived in Harlem, New York, with their three daughters, ages fourteen, twelve, and seven. To celebrate his emerging success, Zachary took his family on their first-ever vacation, to the Seville Hotel on Miami Beach.

The Seville Hotel is situated on the ocean at 29th Street. The City of Miami Beach had equipped the beach behind the Seville Hotel with public restrooms with showers, water fountains, telephones, picnic tables, and metered parking. Additionally, the city issued a license to a concessionaire, Hurricane Beach Rentals, to operate on the beach behind the Seville Hotel, renting a variety of equipment, including lounge chairs, umbrellas, and watercraft. In order to obtain its license from the city, Hurricane Beach Rentals had to obtain written approval of the Seville Hotel, and agreed to pay the hotel a monthly fee in exchange for its consent.

The City of Miami Beach had manned lifeguard stations at various locations along the beach, but there was none at 29th Street. In fact, the nearest lifeguard stands were eight blocks to the south, at 21st Street, and six blocks to the north, at 35th Street. The 29th Street location was the only location on Miami Beach that had public restrooms and other facilities but did not have a lifeguard station.

The Miami Beach lifeguards used a flag system to warn the public of rip currents, but only in locations where there were lifeguard stations. On the day of the accident, the city lifeguards knew that there were active rip currents all along Miami Beach, and rip current warning flags were posted at each lifeguard station. However, there were none at 29th Street.

The Saxony Hotel is situated on the beach some three blocks north of the Seville Hotel. Eugenie Poleyeff and her husband, Rabbi Israel Poleyeff, were guests at the Saxony Hotel. Observing that there were beach lounges, umbrellas, and lots of people behind the Seville Hotel, the Poleyeffs walked three blocks to the south and rented a double beach lounge and umbrella from Hurricane Beach Rentals. At about the same time, the Breaux family also rented a double beach lounge from Hurricane Beach Rentals. Frederica made a video recording of Zachary building a sandcastle with his daughters and wading at the water's edge. Frederica Breaux, Rabbi Poleyeff, and two independent eyewitnesses testified that they believed the Hurricane Beach Rental employees to be lifeguards, and believed that the Hurricane Beach Rental hut was a lifeguard station. They testified that the employees dressed and acted like lifeguards. Their license with the city required them to wear City of Miami Beach badges although there was no testimony that they were wearing them at the time of the accident. There were no warning signs to indicate that the area was not a designated swimming area, that it was unguarded, or where the nearest guarded beaches were. All eyewitnesses testified that a number of people were swimming or wading in the ocean. There was also testimony that the city was aware that the public used the 29th Street beach area for swimming and watercraft rental.

Unaware that there were dangerous rip currents, Eugenie Poleyeff went for a swim in the ocean and began screaming for help as the rip currents began to overtake her. Zachary, building a sandcastle with his daughters at the edge of the beach, pointed at the Hurricane Beach Rentals kiosk and instructed his oldest daughter to get a lifeguard. Zachary then entered the ocean in an attempt to rescue Eugenie Poleyeff, who was a complete stranger to him. The daughter alerted Frederica and the two of them ran to the Hurricane Beach Rentals hut screaming for help. Tragically, there were no lifeguards, only concessionaires who looked like lifeguards. Frederica and her three daughters watched helplessly while both Zachary and Eugenie were overtaken by the rip currents and drowned. When a swimmer was able to pull Zachary's body to the beach, Frederica performed CPR until paramedics arrived while her daughters watched in horror. Unfortunately, it was too late and Zachary was gone.

It is important to note that the beach and ocean are owned by the state of Florida, but were operated by the City of Miami Beach under a contract with the state. The city paid the state a percentage of fees paid by concessionaires. Further, the city advertises nationally and internationally with photos of the beach and ocean in order to attract tourists.

Separate wrongful death lawsuits were filed in Dade County Circuit Court on behalf of the estates of Breaux and Poleyeff against the City of Miami Beach, the Seville Hotel, the Saxony Hotel, and Hurricane Beach Rentals. The claims against the city alleged in part that the city breached its duties to the swimmers because it failed to operate a designated water recreational area in a reasonably safe manner. The city also failed to provide lifeguards to protect users of the beach in an area where the public was invited. Furthermore, the city attracted to use the beach and ocean by virtue of the existence of public facilities, along with the operation of concessions renting beach equipment, watercraft and/or other water recreational equipment. The complaints also alleged that the city failed to have a warning system for rip currents or other dangerous surf conditions. In the alternative, the estates alleged that the city failed to post warnings that the beach was unguarded, and that swimmers should proceed to 21st Street or 35th Street, where lifeguards were present before entering the ocean.

Breaux v. City of Miami Beach, 899 So.2d 1059 (Fla. 2005)

The claims against the hotels and Hurricane Beach Rentals were based on the legal theory that an occupier of premises owes two basic duties to an invitee (*La Villarena, Inc. v. Acosta*, 597 So.2d 336 (3 DCA 1992)):

1. To use ordinary care in keeping premises in a reasonably safe condition.
2. To warn of latent or concealed perils that are known or should be known to the owner or occupant, of which the invitee is not aware, and which the invitee cannot discover through the exercise of reasonable care.

Historically, Florida courts have protected the hotel and tourism industry in such cases based on the legal fiction that the hotels do not own or occupy the beach or ocean, and whether guests swim in the ocean is a matter of individual choice and not intrinsically linked to the hotel's business. See *Adika v. Beekman Towers, Inc.*, 633 So.2d 1170 (3 DCA 1994); *Sperka v. Little Sabine Bay, Inc.*, 642 So.2d 654 (1 DCA 1994).

The plaintiffs attempted to distinguish earlier cases through evidence that Hurricane Beach Rentals actually conducted its business on the beach and in the ocean and that its kiosk was physically on the beach; its employees physically worked on the beach; the lounges, umbrellas, and cabanas it rented were on the beach; and the watercraft it rented were in the ocean. Moreover, Hurricane Beach Rentals was required by the City of Miami Beach to place buoys in the ocean to designate a channel for the watercraft to follow and separate it from a designated swimming area. The plaintiffs further argued that the Seville Hotel exercised actual control over the beach and ocean in that Hurricane Beach Rentals would not have been permitted to operate its concessions on the beach without written permission of the Seville Hotel, for which the hotel received $500 monthly payments. The plaintiffs further alleged that the business of the hotels was intrinsically linked to guests using the beach and swimming in the ocean and that the hotels advertised accordingly.

The plaintiffs relied in part on the case of *McKinney v. Adams*, 68 Fla. 208, 66 Southern 988 (1914). In that case, decided over a century ago, the operator of a seaside bathhouse on a public beach (that among other things rented bathing suits to swimmers) was held negligent for failing to provide lifeguards and safety equipment. The *Adika* court distinguished the *McKinney* case, finding that Beekman Towers was adjacent to but not on the beach, as opposed to the public bathhouse in *McKinney* which "profited from the rental of equipment expressly intended for use while swimming in the adjacent Atlantic Ocean." *Adika v. Beekman Towers, Inc.*, supra at 1171. Although part of the holding in *McKinney* rested on the defendant's failure to comply with a statute that has since been repealed, the factual distinction between an occupier of the beach profiting from rental of equipment expressly intended for water recreational use seemed factually close to the operations being performed by Hurricane Beach Rentals.

Similarly, in *Sperka v. Little Sabine Bay, Inc.*, 642 So.2d 654 (1 DCA 1994), the case turned on the defendant's lack of control over the beach and ocean area. As in *Adika*, the defendant in *Sperka* was a hotel that was not conducting its business on the beach and did not exercise control over the beach. The court in *Sperka* also pointed out that "... a duty to warn against unusual dangers not generally existing in similar bodies of waters arises only when that business enterprise defines or designates in some fashion a specified area for the recreational use of its patrons..." *Id.* at 655. The complaints of Breaux and Poleyeff specifically alleged that Hurricane Beach Rentals "... had designated specific areas for use of its rental equipment at the direction of the City of Miami Beach." The complaints also alleged that the hotels engaged in business activities connected with use of the beach and ocean adjacent to its premises and exercised control or dominion over the beach or ocean by acts which included giving permission to concessionaires and vendors to rent beach lounges, chairs, watercraft, or other water recreation equipment to guests of the hotel and others. The complaints also alleged that by authorizing Hurricane Beach Rentals to conduct business on the beach behind its hotel, the Seville encouraged and invited beach and ocean users, including the plaintiff, to use the beach and ocean behind the Seville Hotel for water recreation purposes.

With regard to negligence allegations, it was alleged that the hotel knew or should have known that the nearest lifeguard towers were at 21st Street and 35th Street and that it was reasonably foreseeable that the City of Miami Beach lifeguards would not be available in the area behind the hotel premises to provide adequate lifeguard protection for swimmers in the area. It was also alleged that the plaintiffs did not know of the dangers of riptide or the existence of rip current in the ocean on Miami Beach on February 20, 1997. The complaints alleged that the rip currents were not visible or obvious hazards to untrained observers and constituted a trap or hidden danger.

The specific negligence allegations against the Seville Hotel included claims that it was reasonably foreseeable to the hotel, based on its advertising the use of the beach and ocean in attracting guests, in providing access from the hotel premises to the beach and in authorizing Hurricane Beach Rentals to conduct its business on the beach; that patrons of the Seville Hotel and others would make use of the beach and ocean behind the hotel premises; and that it was also reasonably foreseeable to the Seville Hotel that, from time to time, dangerous rip currents would be present in the ocean behind their hotel premises and that guests of the hotel would be unfamiliar with the identification of rip currents, the dangers of rip currents, or procedures for escaping danger in the presence of rip currents.

The complaints alleged that the Seville Hotel breached their duties in the following respects:

1. Failed to implement a sufficient safety program for users of the beach and ocean behind the hotel premises.
2. Failed to have a warning system for rip currents or other dangerous surf conditions by the use of warning signs, warning flags, condition boards, or other warning means.
3. Failed to have an educational program for patrons of the hotel, by use of brochures, television access, or other means to provide information to hotel patrons concerning the dangers of rip currents, identification of rip currents, or safety procedures to follow when confronted with rip currents.

4. Failing to have available on their premises and within close proximity to the beach safety equipment such as throw bags with attached lines, life preservers, or a dedicated emergency telephone line.
5. Failing to warn that the beach was unguarded and that swimmers should proceed eight blocks south to 21st Street or six blocks north to 35th Street, where lifeguards were present, before entering the ocean.

A conscious decision was made to avoid any allegation that the hotels, as upland property owners, owed a duty to supply ocean lifeguards. Obviously, the statewide economic impact of requiring the oceanfront property owners to hire and train ocean lifeguards would be enormous. It was thought that the seeking of such a ruling from our courts would be unpalatable and likely to fail. Rather, the thrust of the allegations centered on failing to warn and educate hotel patrons about the dangers and identification of rip currents, how to escape rip currents, the fact that this beach was unguarded, and the location of the nearest lifeguarded beaches.

Consideration was also given, and ultimately rejected, as to alleging that the hotels should have had automated external defibrillators (AEDs) available in reasonable proximity to the beach. However, AED litigation was in its infancy at the time and there were concerns that injecting such a collateral issue into the case could detract from the chance of prevailing on the already difficult duty issues being presented.

In addition to the negligence count, the complaint alleged separate counts under the Rescue Doctrine. It was alleged that the defendant had created a situation of peril for Mrs. Poleyeff and that Zachary Breaux acted reasonably in entering the ocean to attempt her rescue when she was yelling for help and was in imminent peril with no lifeguard in the vicinity. In the unlikely event that there was a judicial finding of a duty owed to Mrs. Poleyeff but not to Mr. Breaux under the negligence count, the Rescue Doctrine could have provided an avenue of recovery for the Breaux estate. It could also have provided a basis for a cause of action for the Breaux estate against the Saxony Hotel, where the Poleyeffs were staying.

The trial court dismissed the complaints against both hotels and Hurricane Beach Rentals with prejudice based on the holding of *Adika v. Beekman Towers, Inc.*, supra, that there is no non-statutory duty imposed upon a beachside hotel to warn its guests against the dangers of a riptide in an adjacent part of the ocean.

Appeal was taken to the Third District Court of Appeals, which granted an en banc hearing and affirmed the trial court dismissal with prejudice. See *Poleyeff v. Seville Beach Hotel Corporation*, 782 So.2d 422 (Fla. Yd DCA 2001), review denied, 817 So.2d 849 (Fla. 2002). The Third DCA specifically endorsed and reaffirmed the holding of *Adika*, holding that an entity which does not control the area or undertake a particular responsibility to do so has no common law duty to warn, correct, or safeguard others from naturally occurring, even if hidden, dangers common to the waters in which they are found (*Adika v. Beekman Towers, Inc.*, supra at 424). The court did not explain why the allegations in the complaints of control over the area by the Seville Hotel and Hurricane Beach Rentals were not legally sufficient to overcome a motion to dismiss. The only comment in the majority opinion about the status of the defendants stated, "… the businesses of operating hotels and renting beach chairs only tangentially or collaterally relate to their customers' use of the ocean." (*Id.* at 424.) The majority opinion summed up the matter as follows:

> It is enough to say that drowning because of a natural characteristic of the very waters in which it occurs is simply one of the perhaps rapidly diminishing set of circumstances for which, without more, no human being or entity should be considered "to blame" deemed "at fault" or, therefore, held civilly liable. (*Id.* at 425.)

The City of Miami Beach filed a motion for summary judgment, arguing that it was entitled to sovereign immunity. The trial court granted the city's motion for summary judgment, finding that the city was immune from suit. Appeal was taken to the Third District, and the district court affirmed the trial court's grant of summary judgment, but not on the basis of sovereign immunity. Instead, relying exclusively on its prior en banc decision in *Poleyeff*, the Third District held that the city had no duty to warn the decedents of, or safeguard them from, the naturally occurring rip currents because it did "not control the area or undertake a particular responsibility to do so." (*Poleyeff v. City of Miami Beach* 818 So.2d 672 (Fla. 3 DCA 2002) (PoleyeffII).) However, in a lengthy dissent, Justice Cope pointed out that based on the evidence presented, it was clear that the City of Miami Beach did control the beach under a lease from the State of Florida. The dissent pointed out that the city exercised its control when it obtained revenue from Hurricane Beach Rentals to which the city granted a license to operate a concession, and by constructing and operating restrooms with showers, drinking fountains, and parking, and by providing access to the beach from its boardwalk. The dissent further argued that the majority ruling was in direct conflict with the Florida Supreme Court opinion of *Florida Department of Natural Resources v. Garcia*, 753 So.2d 72 (Fla. 2000). In that case, the Supreme Court considered the liability of the state for an accident in the same area of South Beach where the Breaux/Poleyeff drownings occurred. The Supreme Court ruled that where an area such as South Beach is a well-known public swimming area from which the state is actively deriving profit, the state has no basis for claiming immunity from suit merely because a formal designation as a state park did not occur (*Id.* at 77).

The dissent cited the Florida Supreme Court as follows:

> A governmental entity that operates a swimming facility "assumes the common law duty to operate the facility safely, just as a private individual is obligated under like circumstances." *Avallone v. Board of County Commissioners*, 467 So.2d 826 (Fla. 5th DCA 1985). Thus, a government entity operating a public swimming area will have the same operational-level duty to invitees as a private landowner—the duty to keep the premises in a reasonably safe condition and to warn the public of any dangerous conditions of which it knew or should have known. See, e.g., *Avallone* 493 So.2d at 1005; *Brightwell v. Deem*, 90 So.2d 320, 322 (Fla. 1956); *Hviazewski v. Wet 'n Wild, Inc.*, 432 So.2d 1371, 1372 (Fla. 5th DCA 1983). (*Id.* at 75)

On March 24, 2005, more than eight years after the drownings, the Florida Supreme Court reversed the Third DCA and held that the City of Miami Beach was operating a public swimming area on the beach and, therefore, had an operational-level duty of care to warn the public of any dangerous conditions of which it knew or should have known (*Breaux v. City of Miami Beach*, 899 So.2d 1059 (Fla. 2005)). The court ruled that the Third DCA's narrow focus on the lack of designation by the Miami Beach City Council of the 29th Street Beach location as a swimming area was not dispositive. Instead, the court directed that all of the circumstances must be considered to determine whether the city was operating a swimming area at 29th Street. The court stated,

> We conclude that the totality of the circumstances in this case demonstrates that the City was operating a "public swimming area" at the 29th Street location. The City knew that the public was using this location for swimming. There were no signs warning the public not to swim and both the Poleyeff family and the Breaux family saw people using the area for swimming. Moreover, although the City did not have a lifeguard station at the 29th Street Beach area, the City built beach facilities at this location and provided metered parking at the end of 29th Street. Of even greater significance, the City licensed a concessionaire to rent beach chairs, umbrellas, and watercraft at this location, thereby deriving revenue from the public's use of this particular beach area. (*Id*. at 1065)

The emphasized sentence in the above paragraph demonstrated the importance to the Supreme Court, in determining a duty owed by the city, of the city having licensed a concessionaire, thereby deriving revenue from the public's use of this particular beach area. This language later would become of paramount importance in the subsequent declaratory judgment lawsuit that will be discussed in the following.

In *Breaux v. City of Miami Beach*, the Supreme Court attempted to define and clarify what had been a confused and unsettled area of the law concerning governmental responsibility for beach and ocean injuries. Some of the important questions which were laid to rest are as follows:

1. A government unit has the discretionary authority to operate or not operate swimming facilities and is immune from suit on that discretionary question. However, once the unit decides to operate the swimming facility, it assumes the common law duty to operate the facility safely, just as a private individual is obligated under like circumstances.
2. The operator of a swimming facility owes a common law duty to keep the premises in a reasonably safe condition and to warn the public of any dangerous conditions of which it knew or should have known.
3. In determining whether a governmental entity is operating a public swimming area, one must examine the totality of the circumstances rather than whether the governmental entity formally "designated" the beach as a public swimming area.
4. The "common use" of an area for swimming may be one factor to consider in determining if a governmental entity held out the area as a public swimming area or lead the public to believe that the area was designated as a swimming area.

5. If the governmental entity held the area out to the public as a swimming area, or led the public to believe the area was a designated swimming area, an operational-level duty of care is owed to those using the swimming area.
6. The duty to keep the premises in a reasonably safe condition applies only to the extent the premises are improved or maintained by the operator. An operator cannot be charged with keeping an unaltered natural body of water "safe." However, the natural character of a hazard does not relieve the operator of a duty to warn if it knew or should have known the hazard was present.
7. The duties of a governmental entity are not limited to manmade dangers. They include natural dangers such as dangerous currents if the operator knew or should have known of the hazard.
8. The fact that rip currents are transient in nature is not dispositive on the issue of duty to warn. The focus is not on the nature of the dangerous condition, but on whether the governmental entity knew or should have known of the dangerous condition.

The Supreme Court quashed the Third DCA decision in *Poleyeff* and remanded the case for further proceedings based on the issue of fact as to whether the City of Miami Beach knew or should have known of the dangerous rip currents at 29th Street on the day of the accident. In doing so, the court expressly disagreed with the Third District's statement that "drowning because of a natural characteristic of the very waters in which it occurs is simply one of the perhaps rapidly diminishing set of circumstances for which, without more, no human being or entity should be ... held civilly liable" (*Id.* at 1065).

While the Supreme Court opined that it was an issue of fact as to whether the city knew or should have known of rip currents at 29th Street on the day of this incident, the court acknowledged that on the day of the accident the lifeguard at 21st Street, eight blocks away, had posted rip current warning flags (*Id.* at 1062). There had also been ample testimony of multiple lifeguards that on the day of the accident all of the lifeguards knew that there were active rip currents all along Miami Beach. Further, it was the policy of the city to post rip current warning flags in all lifeguarded locations if there were rip currents identified at any of the locations. Thus, it was clear that the purported issue of fact would almost certainly result in a jury determination that the city knew or should have known of the rip current hazard on the day in question.

The Supreme Court ruling obviously created a dilemma for the City of Miami Beach. Now that the litigation was reinstated it was clear there was little the city would be able to do to defend liability. Comparative negligence obviously was not a valid defense under the circumstances. The city was entitled to its sovereign immunity defense with a $200,000 cap. However, there was no question that once a judgment was entered against the city following a jury trial, a claims bill would be filed in the Florida Legislature. Testimony years later of Miami Beach City attorneys in a federal court declaratory judgment lawsuit revealed that the city estimated that a jury verdict for the Breaux estate would easily exceed $10,000,000. The city had already spent $327,000 defending the case and did not have the resources in-house to continue to try it on remand. The city estimated that it would cost an additional

$500,000 to continue defending with outside counsel through verdict and appeal. The city attorney was concerned about the precedential effect of a verdict of the magnitude expected on other cases against the city, the negative effect on tourism due to publicity surrounding a significant verdict, and the effect of such a verdict on the city's municipal bond credit rating.

The concessionaire, Hurricane Beach Rentals, had been dismissed from the case and had prevailed on appeal. However, in obtaining its license from the City of Miami Beach to operate behind the Seville Hotel, the city had required that Hurricane Beach Rentals procure a $1,000,000 liability insurance policy naming the city as an additional insured. Counsel for Breaux, Polyeff, and the city attorneys carefully examined the language of the insurance policy which provided as follows:

> In consideration of the additional premium charged, $INCLUDED, it is hereby understood and agreed that the "Persons Insured" provision is amended to include as additional insureds the persons or organizations named below but only with respect to liability arising out of the operations performed for such additional insured by or on behalf of the named insured.

The claimant attorneys and the city attorneys scrutinized the language used by the Florida Supreme Court stating, "Of even greater significance, the City licensed a concessionaire to rent beach chairs, umbrellas, and watercraft at this location, thereby deriving revenue from the public's use of this particular beach area." They strategized that the Supreme Court reliance on the operation of concessions in concluding that the city was operating a "public swimming area" at 29th Street satisfied the insurance policy language. Based on the Supreme Court decision, the liability of the city arose out of Hurricane Beach Rentals having performed operations for the city. Accordingly, Breaux and Polyeff made a joint demand for the $1,000,000 policy limits and the City of Miami Beach demanded that the insurance company acknowledge coverage and provide the city with a defense to the Breaux and Polyeff lawsuits.

At this point the insurance company erroneously reasoned that it could not have coverage for an additional insured when the named insured had been found to have no liability. Thus, the insurance company denied coverage and also denied a defense to the city. The insurance company filed a declaratory judgment lawsuit in federal court, asserting that it did not owe coverage or a defense to the city. Breaux, Polyeff, and the City of Miami Beach filed counterclaims. As in the underlying state court proceedings, Frederica Breaux and her daughters were represented by Howard Pomerantz and appellate counsel Nancy Little Hoffmann. Rabbi Polyeff was represented by Andy Yaffa and appellate counsel Joel Eaton. The city was represented by Jose Smith, Donald Papy, and Judith Weinstein.

Florida Law provides that when an insurer wrongfully denies a defense to its insured, leaving the insured on its own, the insured is entitled to negotiate a settlement with the injured parties and assign its rights to the injured parties to sue the insurer to enforce and collect the settlement amount. Such a settlement agreement is known as a Coblentz Agreement. (See N.Y. 416 F.2d 1059 (5th Cir. 1969).) After extended and complex negotiations and with the blessings of the Miami Beach City Commission, the city entered into settlement agreements with Breaux and Polyeff, totaling $5,000,000 to the Breaux family and $750,000 to Rabbi Polyeff, who had subsequently remarried. The settlement agreements provided for Miami Beach to pay its $200,000 sovereign immunity limits and the claimants agreed not to execute against the city beyond their $200,000 cap. Any additional funds to satisfy the settlements would have to come from the insurer.

In order to enforce a Coblentz Agreement, an injured party must prove three things:

1. Coverage.
2. The insurer wrongfully refused to defend the insured.
3. The settlement is reasonable and was made in good faith.

United Stated District Court Judge Alan S. Gold bifurcated the trial such that the first trial (Phase 1) tried the issues of whether the Hurricane Beach Rental insurance policy provided coverage to the City of Miami Beach as additional insured and whether the insurer had wrongfully refused to defend the city. The second trial (Phase 2) addressed the issues of whether the settlements were reasonable and made in good faith.

The Phase 1 trial resulted in a verdict in favor of Breaux, Polyeff, and the city, holding that coverage existed and the refusal of the insurance company to furnish a defense was wrongful. Phase 2 of the trial presented a number of fascinating factual issues. Frederica Breaux and her daughters asserted that Zachary had emerged as a budding star in the world of jazz and hip hop music. Rickey Schultz, President of Zebra Records, testified as an expert witness to support the economic damage claims of Breaux, which an economist placed between $2,500,000 and $6,200,000, present money value. The insurance company called upon Peter Shukat, a highly regarded entertainment/music industry attorney from New York, who concluded that Zachary would not have been successful in economic terms. Judge Gold accepted the testimony of Rickey Schultz, finding that he had a career of over thirty years in evaluating the probabilities of commercial success of emerging jazz musicians and had signed Zachary Breaux to a six-album contract with his own record company.

A major issue in Phase 2 was whether a claims bill following verdicts in favor of Breaux and Polyeff would have been successfully passed in the Florida legislature. Each side had Tallahassee lobbyists as claims bill experts. Judge Gold ultimately ruled, after the city attorneys testified that they would have joined in a claims bill in order to compromise a large verdict, that the city had a reasonable fear that the Breaux and Polyeff claims bills would have passed the legislature in substantial amounts. Accordingly, Judge Gold ruled that the settlement amounts were reasonable.

In determining whether the settlements were made in good faith, Breaux, Polyeff, and the city presented evidence of their extensive negotiations. Judge Gold ruled that there was no overreaching or collusion involved in the settlements and the city vigorously negotiated the settlements over a period of more than one year. He found that the settlement amounts were reached in good faith.

Final judgments were entered in the sum of $5,939,386 for Breaux, $871,273 for Polyeff, and $245,429 for the City of Miami Beach (which sums included prejudgment interest). The court reserved ruling as to the amount of attorney's fees and costs to be awarded. The total attorneys' fees and costs being sought amounted to more than $3,000,000.

The insurance company appealed to the 11th Circuit Court of Appeals. During the pendency of the appeal, the insurance company settled with Polyeff and the City of Miami Beach for confidential amounts. Just prior to an anticipated ruling by the 11th Circuit Court of Appeals, settlement concluded with Breaux for $5,000,000.

Three months after the Florida Supreme Court handed down its ruling in the Breaux case, the Florida legislature passed Florida Statute 380.276 regarding beaches and coastal areas. Section 6 of this statute legislatively reversed part of what the Florida Supreme Court had established regarding liability of governmental entities in the *Breaux* decision. Section 6 reads as follows:

> Due to the inherent danger of constantly changing surf and other naturally occurring conditions along Florida's coast, the state, state agencies, local and regional government entities or authorities, and their individual employees and agents, shall not be held liable for any injury or loss of life caused by changing surf and other naturally occurring conditions along coastal areas, whether or not uniform warning and safety flags or notification signs developed by the department are displayed or posted.

While this statute will certainly prevent the survivors of many drowning victims from having their day in court and opportunity to be fairly compensated for negligence of governmental entities, the statute, when carefully examined, leaves opportunity under some circumstances for victims to receive some measure of justice. The statute only applies to "... the state, state agencies, local and regional government entities or authorities, and their individual employees and agents ..." Thus, the statute does not apply to private businesses if they are not agents of government entities, such as hotels, condominiums, restaurants, concessionaires, marine construction companies, and so forth. Although the *Breaux* case was unsuccessful in extending liability to the hotels and concessionaire, neither the Third District Court of Appeals or the Florida Supreme Court ever gave an explanation of why they were not liable. Review of the applicable case law would lead one to believe that Hurricane Beach Rentals met the criteria necessary to owe a legal duty to use ordinary care in keeping the premises reasonably safe and to warn of latent or concealed perils.

The injuries that F. S. 380.276(6) apply to are "... any injury or loss of life caused by changing surf and other naturally occurring conditions ..." The statute was obviously intended to apply to rip currents. It would probably also apply to injuries caused by ocean life. However, it would not apply to manmade hazards such as a change in surf conditions caused by beach renourishment. Nor would it apply to an injury caused to a swimmer who struck his head on debris left behind from demolition of a pier as was the case in *Florida Department of Natural Resources v. Garcia*, 753 So.2d 72 (Fla. 2000). Similarly, injuries caused by manmade objects sunken to create artificial reefs would not be prohibited by the statute.

F.S. 380.276(6) applies to injuries that occur "... along coastal areas ..." *Coastal* is defined as the land adjacent to an ocean or sea. Thus, the statute should not apply to any body of water in the State of Florida other than the Atlantic Ocean or the Gulf of Mexico. For instance, the victim of a drowning in Lake Okeechobee would not be prohibited from bringing suit.

The constitutionality of Florida Statute 380.276(6) was upheld by the Fourth District Court of Appeals in the case of *Brown v. City of Vero Beach*, 64 So.3d 172 (4 DCA 2011). In that case, fourteen-year-old Eric Brown drowned trying to save his fifteen-year-old friend, who was caught in an ocean rip current off of South Beach Park in Vero Beach. Their case alleging the city breached its duty to warn the public of dangerous conditions in the ocean was dismissed based on Florida Statute 380.276(6). The court affirmed the dismissal holding that the statute's language is clear and unambiguous. Therefore, the plain and ordinary meaning controls and the courts may not resort to legislative history or other rules of statutory construction to discern its meaning. The plaintiff argued that the amendment abrogated a long-standing common law right. However, the court held that argument was without merit because the waiver of sovereign immunity did not yet exist as of July 4, 1776, or at the time of ratification of the Declaration of Rights in the Florida Constitution. Therefore, the Florida Legislature had the discretion to place limits and conditions on the scope of sovereign immunity waiver. The court cited to the cases of *Campbell v. City of Coral Springs*, 538 So.2d 1373 (4 DCA 1989), and *Cauley v. City of Jacksonville*, 403 So.3d 379 (Fla. 1981).

This author is of the strong opinion that Florida Statute 380.276(6) is an unfair and unnecessary invasion of the constitutional rights of drowning victims and their families. Governmental entities already had the protection of the sovereign immunity laws which currently limit recovery against a governmental entity to $200,000 per person and $300,000 per claim. In the event of negligence on the part of a governmental entity there is no good reason to treat the victims of drownings differently than victims of other types of injuries or deaths attributable to governmental negligence. When Florida Statute 3 80.276 was amended in 2005 to provide complete governmental immunity in this instance, there was no floor debate whatsoever in either the House or Senate and the bill was passed in each house unanimously. It does not appear that most legislatures were even aware that they were voting for complete governmental immunity at the expense of drowning victims when sovereign immunity laws already granted adequate protection to the governmental coffers. It appears that the only way to correct this injustice will be through legislative repeal of F.S. 380.276(6).

This author would be remiss in failing to point out that no part of any success achieved in handling the *Breaux* case would have been possible without the extraordinary efforts of appellate attorney Nancy Little Hoffmann, who steered the journey through the myriad legal issues involved in these complex cases at both the trial and appellate levels. As a final note, the old adage "every cloud has a silver lining" may apply to the *Breaux* case. Possibly as a result of more than a decade that it took for the *Breaux* case to litigate its way through the state and federal courts, including multiple appeals, one of Zachary and Frederica Breaux's daughters developed an interest in the law and is now a practicing attorney in the state of Texas.

10 Past Its Prime
The 1920 Death on the High Seas Act

Michael D. Eriksen[1]

CONTENTS

General Maritime Jurisdiction and DOHSA Jurisdiction: An Incomplete Overlap	211
The Calhoun Case	213
TWA Flight 800	213
The *Deepwater Horizon*	213
Pre-DOHSA Claims in the United States for Maritime Wrongful Death	214
The *Harrisburg*	214
Other Practical Limits on Tort Claims in the Late 1800s	215
The Enactment of DOHSA (1903–1920)	215
What Are the "High Seas"?	216
"High Seas" in International Law	216
"High Seas" in DOHSA's Legislative History	217
"High Seas" in Case Law	217
DOHSA Going Forward	218
Congress Should Fix DOHSA	218
Courts Can and Should Limit DOHSA's Geographic Application	219
Endnotes	220

> Certainly it better becomes the human and liberal character of proceedings in admiralty to give than to withhold the remedy, when not required to withhold it by established and inflexible rules.
>
> *The Sea Gull*, 21 F. Cas. 909, 910 (C.C. Md. 1865) (Chase, C.J.)

Maritime law is one of the most complex areas of American law. No aspect is more tangled than the remedies for maritime wrongful deaths of "non-seafarers," that is, those who are not seamen or longshore workers.

Depending on the fortuity of where a non-seafarer's fatal injury occurs on *or in* the world's navigable waters, his or her survivors who sue in courts in the United States either may be able to invoke modern maritime choice of law rules to access economic *and non-economic* damages under state wrongful death laws or they may be limited to their "pecuniary" losses (i.e., economic damages) by the 1920 Death on the High Seas Act ("DOHSA").

In 1920, state wrongful death laws generally provided only economic damages, like DOHSA. As time passed, however, most states added *non-economic* compensatory wrongful death elements of damage which recognize that human beings are more than economic assets to their families.[2] Additionally, the Supreme Court has acknowledged that general maritime common law allows non-pecuniary *punitive* damages in tort cases.[3] DOHSA has not kept pace with these developments. As a result, the pecuniary damages allowed by DOHSA have become a shield for marine tortfeasors rather than the claimants' sword they once were.

This chapter puts DOHSA in historical context, and argues for parity between federal maritime wrongful death elements of damage and those prevailing in the states. That *was* the situation when DOHSA was passed in 1920, but is not now.

DOHSA, as recodified (i.e., renumbered) by Congress in 2006, today reads:

TITLE 46—SHIPPING
Subtitle III—Maritime Liability
CHAPTER 303—DEATH ON THE HIGH SEAS

Sec. 30302. Cause of action.

When the death of an individual is caused by **wrongful act, neglect, or default occurring on the high seas beyond 3 nautical miles from the shore of the United States**, the personal representative of the decedent may bring a civil action in admiralty against the person or vessel responsible. The action shall be for the exclusive benefit of the decedent's spouse, parent, child, or dependent relative. (emphasis added)

Sec. 30303. Amount and apportionment of recovery.

The recovery in an action under this chapter shall be *a fair compensation for the pecuniary loss* sustained by the individuals for whose benefit the action is brought. The court shall apportion the recovery among those individuals in proportion to the loss each has sustained. (emphasis added)

Sec. 30304. Contributory negligence.

In an action under this chapter, contributory negligence of the decedent is not a bar to recovery. The court shall consider the degree of negligence of the decedent and reduce the recovery accordingly.

Sec. 30305. Death of plaintiff in pending action.

If a civil action in admiralty is pending in a court of the United States to recover for personal injury caused by wrongful act, neglect, or default described in section 30302 of this title, and the individual dies during the action as a result of the wrongful act, neglect, or default, the personal representative of the decedent may be substituted as the plaintiff and the action may proceed under this chapter for the recovery authorized by this chapter.

Sec. 30306. Foreign cause of action.

When a cause of action exists under the law of a foreign country for death by wrongful act, neglect, or default on the high seas, a civil action in admiralty may be brought in a court of the United States based on the foreign cause of action, without abatement of the amount for which recovery is authorized.

Sec. 30307. Commercial aviation accidents.

(a) Definition—In this section, the term "nonpecuniary damages" means damages for loss of care, comfort, and companionship.
(b) Beyond 12 Nautical Miles—In an action under this chapter, if the death resulted from a commercial aviation accident occurring on the high seas beyond 12 nautical miles from the shore of the United States, additional compensation is recoverable for nonpecuniary damages, but punitive damages are not recoverable.
(c) Within 12 Nautical Miles—This chapter does not apply if the death resulted from a commercial aviation accident occurring on the high seas 12 nautical miles or less from the shore of the United States.

Sec. 30308. Nonapplication

(a) State Law—This chapter does not affect the law of a State regulating the right to recover for death.
(b) Internal Waters—This chapter does not apply to the Great Lakes or waters within the territorial limits of a State.

DOHSA claims are subject to the three-year federal maritime tort statute of limitations, 46 U.S.C. §30106 (unless a contract such as a cruise ticket specifies a shorter time). DOHSA suits may be dismissed for *forum non conveniens* or for lack of personal jurisdiction, which is determined under applicable long-arm and federal due process requirements.[4]

GENERAL MARITIME JURISDICTION AND DOHSA JURISDICTION: AN INCOMPLETE OVERLAP

Historically, the basic building blocks for any successful tort action have remained constant, wrongful death and maritime cases being no exception. The court must have "jurisdiction over the subject matter." There must be a choice of substantive law that recognizes a cause of action and elements of damage. There must be "personal jurisdiction" over the defendant or *res*, and there must be a reasonable likelihood that any judgment be collectible.

Article III, § 2 of the U.S. Constitution extends federal admiralty judicial power to "all cases of admiralty and maritime jurisdiction." The Judiciary Act of 1789, ch. 20, 1 Stat. 73, conferred original maritime subject matter jurisdiction on federal courts. In the same law, Congress "saved to suitors" the traditional right to pursue *in personam* maritime tort and contract actions *in state court*. Thus, *in personam* maritime tort claimants may choose a state or federal forum. Actions *in rem* against vessels, however, may be brought only in a federal district court sitting in admiralty. Maritime tort suits filed in state court, including DOHSA claims, are not removable to federal court except on diversity grounds.[5]

"The fundamental interest giving rise to [general] maritime [subject matter] jurisdiction is 'the protection of maritime commerce.'"[6] Therefore, the traditional locality test for such jurisdiction has given way to a test which focuses on both *location* and *connection* with maritime activity.[7] The "location" part of the test requires a court to determine whether the tort occurred on navigable waters or was caused by a vessel on navigable water. The "connection" part of the test raises two issues. The first issue is whether the incident has "a potentially disruptive impact on maritime commerce." The second issue is whether "the general character" of the activity giving rise to the incident shows a "substantial relationship to traditional maritime activity." The collision of two pleasure boats, for example, may satisfy these requirements.[8]

"Substantive maritime law" is an amalgamation of federal maritime legislation and general maritime common law (including choice of law rules). The application of substantive maritime law to a state or federal tort claim is required, and is allowed *only*, when the tort is maritime in nature.[9]

"The shore is now an artificial place to draw a line. Maritime commerce has evolved along with the nature of transportation and is often inseparable from some land-based obligations."[10] Therefore, certain land activities such as cruise shore excursions are now subject to maritime subject matter jurisdiction and law.[11]

DOHSA, on the other hand, applies *only* if certain events occur on the high seas (e.g., a "wrongful act, neglect, or default" resulting in death or fatal injury).[12] Location is the *sole* DOHSA jurisdictional inquiry. For example, a recreational swimming or diving drowning in non–U.S waters can theoretically support a DOHSA claim in state or federal court (if personal jurisdiction exists over a defendant), even though not otherwise amounting to "traditional maritime activity" sufficient to invoke regular maritime subject matter jurisdiction.[13]

Thus, a separate DOHSA count *should always be included* in a lawsuit for a death occurring in non–U.S. waters, even if an alternative effort is being made to expand damages by invoking a U.S. state or foreign wrongful death act; and such a suit must be filed *within the shorter of* the DOHSA time limit or that of the alternative wrongful death statute (unless suit limits have been further shortened by contract, e.g., a cruise ticket).

The Calhoun Case

In 1989, a twelve-year-old Pennsylvania girl, Natalie Calhoun, was killed in a Jet Ski accident in the territorial waters of Puerto Rico while vacationing at a resort there. Natalie's parents sued the Jet Ski manufacturer for product liability in federal court in Pennsylvania, which had long-arm jurisdiction over the defendant. DOHSA did not apply because the incident occurred within three nautical miles of the shoreline of a U.S. state or territory.

Because the incident involved a "vessel" and occurred in navigable waters, however, general maritime subject matter jurisdiction existed. Accordingly, in *Calhoun v. Yamaha*, 216 F.3d 338 (3rd Cir. 2000) ("*Calhoun II*"), the Third Circuit U.S. Court of Appeals turned to modern federal maritime common law choice of law rules, including the doctrine of *depecage. Depecage* is the application of the laws of different sovereigns to separate issues in a legal dispute, that is, "choice of law on an issue-by-issue basis."[14]

The Third U.S. Circuit Court of Appeals decided that the basic, underlying liability issues would be governed by substantive general (federal) maritime law to maintain uniform national standards of maritime behavior. The court held that Puerto Rico, where the incident occurred, had the greatest interest in having its law apply to the punitive damages claim. However, because Natalie's survivors would experience their personal harms *where they lived*, the court decided that compensatory damages would be determined under Pennsylvania's wrongful death law, which included certain non-economic damage elements. The Supreme Court had previously held in *Yamaha v. Calhoun*, 516 U.S. 199, 210–11, 213 (1996) ("*Calhoun I*") that the national maritime uniformity principle is *not* offended if damages vary depending on which state's wrongful death act is invoked.

Had DOHSA applied, Natalie Calhoun's life would have had little legal value because she was not a wage earner and non-economic damages would not have been allowed.

TWA Flight 800

On July 17, 1996, TWA Flight 800 crashed into the Atlantic Ocean, killing all on board. Because the plane went down approximately nine miles offshore of Long Island, New York, DOHSA applied. That made the lives of sixteen teenaged victims from Pennsylvania nearly worthless from a legal standpoint. To change that outcome, congressional representatives from that state introduced DOHSA Section 30307, which retroactively allowed compensatory non-economic damages for "commercial aviation accidents."[15] However, DOHSA's original non-economic damage prohibition was left intact for all other maritime fatalities, due to intense lobbying on behalf of foreign cruise lines based in Florida.

The Deepwater Horizon

On April 20, 2010, the contemporary relative harshness of DOHSA's pecuniary damages regime was once again brought into stark relief. The *Deepwater Horizon* floating oil platform exploded more than 200 miles offshore of the United States, killing eleven workers and spewing oil into the Gulf of Mexico. Long-arm jurisdiction over the tortfeasors was possible in U.S. states that provide non-economic wrongful death damages to decedents' survivors. However, DOHSA potentially stood in the way. To add insult to injury, the owner of the *Deepwater Horizon* was expected to petition a federal district court sitting in admiralty to limit its liability to the vessel's post-casualty value (like the owner of the *Titanic* did a century ago). An outraged U.S. House of Representatives quickly passed H.R. 5503, which retroactively expanded DOHSA damages and repealed the current Limitation of Liability Act. These measures died in the U.S. Senate after July 15, 2010, when the oil stopped flowing (and in the face of another powerful lobbying effort on behalf of foreign cruise interests).

PRE-DOHSA CLAIMS IN THE UNITED STATES FOR MARITIME WRONGFUL DEATH

Historically, there was no common law right of action for wrongful death in either British or U.S. courts. Such claims were considered to be personal to the decedent rather than to the survivors, and were therefore extinguished by death. That situation was remedied in Britain by the passage of Lord Campbell's Act in 1846.[16] That statute removed common law barriers to *the decedents' survivors* recovering their pecuniary (i.e., economic) losses caused by a wrongful death. Lord Campbell's Act did not extend to *in rem* actions, however. That limited the act's usefulness regarding vessel-related maritime deaths.

By the late 1800s, many U.S. states had passed versions of Lord Campbell's Act, which allowed only economic damages. However, the early state statutes were inconsistent.[17] At least one required a criminal conviction of the tortfeasor as a condition precedent to civil liability. Some were expressly limited to deaths occurring in the state's own territory. Others excluded *in rem* actions altogether.

By contrast, the modern wrongful death acts of U.S. states generally are not so limited; and, as previously mentioned, most (but not all) now provide non-economic compensatory damage elements which recognize human beings as more than economic assets.

In the late 1800s, as now, the high seas were viewed as part of the global commons, that is, an area outside the control of any sovereign.[18] Nevertheless, state and federal courts sometimes applied early state wrongful death laws to high seas fatalities.[19] That would most often occur when the vessels were connected to the states in question, or when the owners submitted themselves and their vessels to federal admiralty court to limit liability for a marine casualty under the federal Shipowners Limitation of Liability Act of 1851.

THE *HARRISBURG*

The *Harrisburg*, 119 U.S. 199 (1886), exposed a so-called void in available maritime death remedies in U.S. courts. The plaintiff's husband was killed when the M/S *Harrisburg*, a steamer from Philadelphia, collided with the decedent's schooner in

Massachusetts waters. The plaintiff was time-barred from suing the at-fault vessel operator in personam under the Massachusetts or Pennsylvania wrongful death statutes. She therefore sued and arrested the vessel in rem in federal admiralty court. The Supreme Court held that there was no distinct general maritime right of action for wrongful death to propel the plaintiff's in rem action (199 U.S. at 219).

The void, if any, was a by-product of the era's prevailing common law choice of law rule, *lex loci delicti*. Under that rigid rule, "the law of the place where the tort was committed" was the *only* choice. However, the high seas lacked a sovereign to provide law for courts to choose. A U.S. maritime wrongful death statute for use in U.S. courts could create the necessary *lex loci*.

Ultimately, the archaic *lex loci deliciti* rule was abandoned for most maritime tort purposes in *Lauritzen v. Larsen*, 345 U.S. 571 (1953). The flexible *Lauritzen* maritime choice of law approach analyzes which sovereigns or states have the most significant relationships to the incident and parties, and the dominant interests in having their law applied. This "most significant relationship" standard (which was applied by the court in *Calhoun II,* supra) has replaced *lex loci delicti* in most other jurisdictions in the United States.[20]

Had the *Lauritzen* choice of law rule been in place before 1920, with *situs* no longer being dispositive, maritime claimants and courts could have more easily invoked the wrongful death acts of states having an interest. That would have significantly reduced the need, if any, for a dedicated federal maritime wrongful death statute.

OTHER PRACTICAL LIMITS ON TORT CLAIMS IN THE LATE 1800S

The broad extraterritorial long-arm jurisdiction we have today was essentially nonexistent at the time of the *Harrisburg*, and thereafter.[21] In that era, the defendant or *res* in dispute in U.S. civil proceedings usually had to be physically served with process in the state where the court sat. That was commonly referred to as *tag jurisdiction*.

Adequate sources of financial indemnity for marine tort liabilities were also far less prevalent than today. A properly arrested vessel could provide both jurisdiction over the *res* and a guarantee of some judgment collectibility. Thus, actions *in rem* against vessels were then a much more valuable and frequently used legal tool in maritime tort cases than now. The *Harrisburg* upset the system in 1886 more than it ever could or would have today.

THE ENACTMENT OF DOHSA (1903–1920)

In retrospect, the *Harrisburg* seems more about the consequences of missing a statute of limitation regarding an existing state wrongful death remedy than demonstrating a need for a new federal maritime wrongful death remedy. Nevertheless, in 1903 the Maritime Law Association of the United States (MLAUS)—then a group of leading commercial maritime lawyers—began to propose specific bills to Congress to create federal maritime wrongful death *lex loci* for use in courts in the United States.[22]

Against this backdrop, the *RMS Titanic* foundered in the North Atlantic on April 14, 1912. Over 1500 people drowned. The British owner was ultimately able to petition a federal court in the United States to limit its liability (to the value of the fares and lifeboats) under the federal Shipowner's Limitation of Liability Act of 1851.[23] The decedents' survivors were denied an adequate remedy, which provided additional political impetus for the passage of a federal maritime wrongful death statute.[24]

The congressional debate on DOHSA centered on whether such a federal statute would and should displace otherwise available state wrongful death acts. That was a classic federal power versus states' rights struggle. At the end, the states' rights advocates prevailed. The final DOHSA in 1920 included the last-minute Mann Amendment, which struck out language expressly limiting state wrongful death statutes to "causes of action accruing within the territorial limits of any State."[25] The Mann Amendment was intended to allow survivors of high seas decedents to elect between DOHSA and state wrongful death laws.

In *Offshore Logistics, Inc. v. Tallentire*, 477 U.S. 207 (1986), however, the Court voted 5-4 to disregard the Mann Amendment for the sake of national maritime uniformity. The majority simply declared that DOHSA *preempts the field* of remedies for fatal "high seas" events (477 U.S. at 227). After *Tallentire*, where DOHSA applies (i.e., in its *locus*) it applies absolutely. In such cases, DOHSA's pecuniary damages elements may not be "supplemented" by either state or foreign non-economic elements, if any.[26]

What Are the "High Seas"?

"Congress confined DOHSA to the high seas."[27] However, the term "high seas" in DOHSA is not formally defined; the Supreme Court has not yet decided if the sovereign waters of foreign countries are included. This is important, because if the relevant fatal events occur entirely outside of DOHSA's locus (i.e., the high seas), but general maritime subject matter jurisdiction otherwise exists, then state wrongful death damages should be accessible as in *Calhoun* II.

"High Seas" in International Law

Before 1492, the high seas were *not* universally recognized to have the same type of geopolitical boundary all the way around. For example, some Europeans reportedly believed that the Atlantic Ocean dropped off the face of the earth west of Portugal.

Pre-DOHSA courts in the United States, on the other hand, distinguished the non-sovereign high seas from the sovereign territorial seas of the (round) world's maritime countries.[28]

The *high seas* are defined in the 1958 Convention on the High Seas, April 29, 1958, 13 U.S.T. 2312 as "all parts of the sea that are not included in the '*territorial sea*' or in the '*internal waters*' of a State" (Art. 1). A maritime country's *territorial seas* are defined in the 1958 Convention on the Territorial Sea and the Contiguous Zone, April 29, 1958, 15 U.S.T. 1607 as "a belt of sea adjacent to its coast" (Art.1). "Waters on the landward side of the baseline of the territorial sea form part of the *internal waters* of the State" (Art. 5) (emphasis added). Both treaties were ratified by the U.S. Senate.

Where, as in DOHSA, such a term has been left undefined in a domestic federal statute, the Supreme Court has readily borrowed the term's formal definition in a Senate-ratified treaty, even one arising after the statute in question. For example, in 1965 the Court did so regarding the undefined key term internal waters in the 1953 Submerged Lands Act.[29] The Court rejected concerns that co-opting such a treaty definition into a domestic statute would impart an "ambulatory quality" to the term, based upon "future changes in international law or practice."

"High Seas" in DOHSA's Legislative History

Throughout the ongoing DOHSA hearings and debates, the term "high seas" was repeatedly given its international, non-sovereign meaning. For instance, in 1912 an MLAUS lawyer stated at a House hearing, "*We leave out the territorial waters of foreign countries.*"[30]

In 1916, the phrase "beyond a marine league (i.e., three nautical miles) from the shore of any State…" was added after "on the high seas" to emphasize that the territorial seas and navigable internal waters of the U.S. were excluded.

DOHSA's title from 1912 until shortly before enactment in 1920 was "A Bill relating to the maintenance of actions for death on the high seas *and other navigable waters.*" At the end, however, the title was narrowed to cover only "death on the high seas," a strong indicator of DOHSA's limited geographical reach.

"High Seas" in Case Law

In *United States v. Louisiana*, 394 U.S. 11, 22 (1969) the Court recognized:

> Nearest to the nation's shores are its *inland* or *internal waters*. These are subject to the complete sovereignty of the nation, as much as if they were a part of its land territory, and the coastal nation has the privilege even to exclude foreign vessels altogether. Beyond the inland waters, and measured from their seaward edge, is a belt known as the marginal or *territorial sea*. Within it the coastal nation may exercise extensive control but cannot deny the right of innocent passage to foreign nations. *Outside the territorial sea are the* **high seas**, *which are in international waters not subject to the dominion of any single nation.* (emphasis added)

In 2000, the Second Circuit U.S. Court of Appeals affirmed a district court's conclusion that the DOHSA phrases "on the high seas" and "beyond three nautical miles from the shore of the United States" do *not* mean the same thing:

> The most natural reading of this text is that a death must occur both on the high seas *and* beyond a marine league from the shore for DOHSA to apply. If a death occurred (1) neither on the high seas nor beyond a marine league; (2) on the high seas but not beyond a marine league; *or* (3) *beyond a marine league, but not on the high seas*, then DOHSA does not apply. (emphasis added)

In Re: Air Crash Off Long Island, New York, on July 17, 1996, 1998 WL 292333, *3 (S.D.N.Y., June 2, 1998), *aff'd*, 209 F.2d 200 (2nd Cir. 2000). The court believed that depriving "high seas" of its own distinct meaning in DOHSA would violate familiar rules of statutory interpretation.

Nevertheless, some Florida U.S. district courts and the Ninth Circuit U.S. Court of Appeals have envisioned a high seas for DOHSA purposes which has only a single boundary (the one closest to home) and which, at its other *unlimited* extremes, may even encompass navigable lakes, rivers, and creeks **within the land-mass** *of other sovereign countries.*

These courts were all influenced heavily by *Sanchez v. Loffland Brothers Co.*, 626 F.2d 1228 (5th Cir. 1980), a Fifth U.S. Circuit Court of Appeals case which involved the death of *a seaman* on navigable Lake Maracaibo in Venezuela.[31] The only issue decided in *Sanchez*, however, was that all of the decedent's survivors' *potential* death remedies (which *possibly* included DOHSA) were time-barred. A single footnote in *Sanchez* observes in passing that a few courts had applied DOHSA to "foreign territorial waters." The issues presented in this article, such as the Senate-ratified treaty definition of "high seas," were not engaged in *Sanchez* because there was no reason to do so.

Three years later, in *Chick Kam Koo v. Exxon Corp.*, 699 F.2d 693, 694–95 (5th Cir. 1983), a different panel of judges of the same Fifth Circuit Court of Appeals determined that a district court's finding, "that DOHSA 'by its title and by its terms' applied only to accidents occurring 'on the high seas' and not to deaths that 'occurred in [the] territorial waters of Singapore,'" **did not** *"overlook controlling statute or case law."* (emphasis added). In other words, according to the panel of judges in *Chick Kam Koo*, DOHSA *does not* apply inside foreign territorial waters (that are the *situs* of most international drownings not connected to major marine casualties like the sinking of the *Titanic*).

With great respect, courts that have over-read *Sanchez* should bravely admit the error, correct it, and move on. Other courts should not compound the error. As the Supreme Court has aptly observed, "[t]he demand for tidy rules can go too far."[32]

DOHSA Going Forward

Had modern, expanded state wrongful death acts, maritime choice of law rules, and long-arm provisions been in place before 1920, DOHSA probably would not have been needed. Additionally, in 1970 the Supreme Court fashioned a general maritime common law right of action for wrongful death, overruling the *Harrisburg* (an original catalyst for DOHSA).[33]

DOHSA is obsolete.

DOHSA today is invoked mostly *by marine tortfeasors*, as a convenient escape hatch from any responsibility for the severe emotional pain and suffering their victims' survivors almost invariably suffer. This is ironic, because DOHSA was enacted to create a remedy, not block them.

Congress Should Fix DOHSA

DOHSA should be amended (1) to allow the possibility of non-pecuniary compensatory and punitive damages in all high seas death cases, and/or (2) to overrule Tallentire by reaffirming Congress' intention, expressed in the 1920 Mann

Amendment to DOHSA, to allow survivors of high seas decedents to elect more generous state wrongful death remedies in lieu of DOHSA in a state court that can assert personal jurisdiction over the wrongdoers.

Fixing DOHSA is urgent, given the high numbers of U.S. non-seafarers who perish annually on non–U.S. navigable waters during vacations, cruises, other trips, or while working. The repair need not wait until the next catastrophic *Deepwater Horizon* or *Costa Concordia*-type disaster occurs just outside our three-mile limit.

Alternatively, DOHSA could simply be repealed. Removing DOHSA as a preempting obstacle in courts in the United States would unleash practical *and more complete* contemporary remedies for maritime wrongful death.

In short (and to paraphrase a salty Marine gunnery sergeant), Congress needs to "lead, follow [the states], or get [DOHSA] the hell out of the way" now.

Courts Can and Should Limit DOHSA's Geographic Application

DOHSA, despite its preemptive quality after Tallentire, was never intended to apply everywhere. The "high seas" simply do not include non-navigable waters, or the sovereign territorial seas and navigable internal waters of the United States and foreign countries.

This conclusion is supported by DOHSA's legislative history, by applicable statutory interpretation canons, by formal definitions of relevant terms in subsequent Senate-ratified treaties, and by official U.S. State Department pronouncements recognizing the internationally accepted outer boundaries of the territorial seas of the world's maritime countries.[34]

As the Court observed in *Lauritzen v. Larsen*, 345 U.S. 571, 577 (1953):

> The shipping laws of the United States, set forth in *Title 46* of the United States Code, comprise a patchwork of separate enactments, some tracing far back in our history and many designed for particular emergencies. While some have been specific in application to foreign shipping and others have been confined to American shipping, many give no evidence that Congress addressed itself to their foreign application and are in general terms which leave their application to be judicially determined from context and circumstance. *By usage as old as the Nation, such statutes have been construed to apply only to areas and transactions in which American law would be considered operative under prevalent doctrines of international law.* (emphasis added)

Where DOHSA does not apply but maritime subject matter jurisdiction and general maritime law do apply, modern common law choice of law rules and *depecage* may be invoked. These rules offer U.S. survivors of non-seafaring maritime decedents a proven route to state wrongful death acts, including any non-economic elements of damage.[35]

The Supreme Court has repeatedly acknowledged that variations in wrongful death *damages elements* (depending on which sovereign's law is applied) do not offend the national maritime uniformity principle, which focuses instead on rules of *maritime behavior*.[36]

Besides, what could be *more* uniform than the term "high seas" having the same meaning throughout the law of the land, particularly in all national statutes and treaties?

ENDNOTES

1. Principal Attorney, Eriksen Law Firm, West Palm Beach, FL; Admiralty Proctor, Maritime Law Association of the United States. One of six Florida attorneys who are currently board-certified by the Florida Bar in both Civil Trial Law and Admiralty and Maritime Law; B.S. Foreign Service, Georgetown University, 1972; United States Marine officer, 1972–77; Univ. of Florida College of Law, 1980 (*Law Review* editor).
2. Carney, C., and Schap, D. Recoverable damages for wrongful death in the states: A decennial view. *J. of Bus. Valuation and Econ. Loss Analysis*, vol. 3, issue 1, article 2 (2008): 1–9.
3. *Atlantic Sounding Co. v. Townsend*, 557 U.S. 404, 411–12 (2009).
4. See, e.g., *Loya v. Starwood Hotels & Resorts Worldwide, Inc.*, 583 F.3d 656 (9th Cir. 2009) (dismissing DOHSA claim for forum non conveniens); and *Fraser v. Smith*, 594 F.3d 842 (11th Cir. 2010) (dismissing DOHSA claim for lack of personal jurisdiction under Florida Long-arm Act).
5. *Romero v. International Terminal Operating Co.*, 358 U.S. 354, 371 (1959); see, e.g., *Argandona v. Lloyd's Registry of Shipping*, 804 F.Supp. 326 (S.D. Fla. 1992) (DOHSA claims filed in state court not removable under federal question jurisdiction).
6. *Norfolk So. Ry. Co. v. Kirby*, 543 U.S. 14, 25 (2004).
7. *Executive Jet Aviation, Inc. v. City of Cleveland*, 409 U.S. 249 (1972); *Sisson v. Ruby*, 497 U.S. 358 (1990); *Jerome B. Grubart, Inc. v. Great Lakes Dredge & Dock Co.*, 513 U.S. 527 (1995).
8. See *Foremost Ins. Co. v. Richardson*, 457 U.S. 668 (1982).
9. *Doe v. Celebrity Cruises, Inc.*, 394 F.3d 891, 899–900 (11th Cir. 2004).
10. *Norfolk So. Ry. Co. v. Kirby*, 543 U.S. 14, 25 (2004); see also, generally, the Extension of Admiralty Jurisdiction Act of 1948, 46 U.S.C. §30101.
11. See, e.g., *Gentry v. Carnival Corp.*, 2011 WL 4737062 (S.D. Fla., Oct. 5, 2011) (cruise excursion).
12. See *Motts v. M/V Green Wave*, 210 F.3d 565, 571 (5th Cir. 2000).
13. See *Delgado v. Reef Resort Limited*, 64 F.3d 642 (5th Cir. 2000).
14. *Black's Law Dictionary* 503 (9th ed. 2009).
15. Cong. Rec., Vol. 143, No. 87, E1282 (June 20, 1997).
16. The *Vera Cruz*, 10 App. Cas. 59 (House of Lords, 1884).
17. H. Rep. No. 1419, "Actions for Death on the High Seas," 64th Cong., 2d Sess., 3 (1917); see also, e.g., the *Alaska*, 130 U.S. 201, 209 (1889) (no *in rem* remedy under New York death act).
18. See e.g., the *Scotia*, 81 U.S. 170, 187 (1871) (the high seas are a place where "no statute of one or two nations can create obligations for the world."); The Scotland, 105 U.S. 24, 29 (1881) (same); The Belgenland, 114 U.S. 355, 369 (1885) (same).
19. See the *Hamilton*, 207 U.S. 398 (1907).
20. See, e.g., *Piamba Cortez v. American Airlines, Inc.*, 177 F.3d 1272 (11th Cir. 1999) (Florida death act applied to Columbia plane crash, under Florida choice of law rule and depecage).
21. See *Pennoyer v. Neff*, 95 U.S. 714 (1878).
22. Whitelock, G. A new development in the application of extraterritorial law to extraterritorial marine torts, 23 *Harv. L. Rev.* 403, 416 (1909).
23. See *Oceanic Steam Nav. Co. v. Mellor*, 233 U.S. 718 (1914); and Eaton and Haas, *Titanic: Triumph and Tragedy*, 297–310 (Haynes UK, 3rd ed. 2011).
24. 51 Cong. Rec. H1928 (1914).
25. 66 Cong. Rec. H4482-87 (1920); see 46 U.S.C.A. § 30308(a) (2006 recodification of DOHSA section which incorporates the Mann Amendment).

26. See, e.g., *Dooley v. Korean Air Lines Co., Ltd.*, 117 F.3d 1477 (D.C. Cir. 1997), aff'd, 524 U.S. 11 (1998) (Korean damages could not supplement DOHSA damages); also, cf., *Mobil Oil Corp. v. Higginbotham*, 436 U.S. 618, 625 (1978) ("Congress did not limit DOHSA beneficiaries to recovery of their pecuniary losses in order to encourage the creation of non-pecuniary supplements.").
27. *Higginbotham*, 436 U.S. at 622.
28. See supra note 17.
29. *United States v. California*, 381 U.S. 139, 166 (1965); also cf., In re Air Crash Off Long Island, New York on July 17, 1996, 1998 WL 292333 *7 (S.D.N.Y., June 2, 1998).
30. Hearing, "Actions for Death on the High Seas," 10 (House Jud. Cmte., August 6, 1912).
31. See, e.g., *Howard v. Crystal Cruises, Inc.*, 41 F.3d 527, 529 (9th Cir. 1994) (citing Sanchez for proposition that DOHSA covers foreign "territorial waters"); and *Moyer v. Rederi*, 645 F. Supp. 620, 623–24 (S.D. Fla. 1986) (same); accord, *Motts v. M/V Green Wave*, 210 F.3d 565, 571 (5th Cir. 2000).
32. *Sisson v. Ruby*, 497 U.S. 358, 364, n. 2 (1990).
33. *Moragne v. States Marine Lines, Inc.*, 398 U.S. 375 (1970).
34. See "Limits In The Seas" series of the Office of Ocean Affairs, U.S. Department of State, http://www.state.gov/e/oes/ocns/opa/c16065.htm.
35. See, e.g., *Calhoun v. Yamaha Motor Corp.*, 216 F.3d 338 (3rd Cir. 2000) ("Calhoun II") (applying modern maritime choice of law rules and depecage); and "In Re: Air Crash at Belle Harbor, New York on Nov. 12, 2001," 2006 WL 1288298 (S.D.N.Y., May 9, 2001) (same; and adopting the most generous available remedy); also cf., e.g., *Piamba Cortez v. American Airlines, Inc.*, 177 F.3d 1272 (11th Cir. 1999) (Florida death act and non-economic damages applied to Columbia plane crash, under Florida choice of law rules and depecage).
36. See *Sea-Land Services, Inc. v. Gaudet*, 414 U.S. 573, 588 n. 22 (1974) ("Congress [in excluding state waters from DOHSA] was not concerned that there be a uniform measure of damages for wrongful deaths occurring within admiralty's jurisdiction."); *Mobil Oil Corp. v. Higginbotham*, 436 U.S. 618, 624 (1978) (same); *Yamaha Motor Corp. v. Calhoun*, 516 U.S. 199, 210–11, 213 (1996) (variations in damages, depending on which state's wrongful death act is applied, do not disturb uniformity of national maritime law).

11 Justice for Deaths Caused by Free-Living Amebae in Recreational Waters

Roger W. Strassburg, Jr.[1]

CONTENTS

Naegleria fowleri: Leading Cause of Death from Recreational Exposure
to Untreated Fresh Water in the United States 224
Duties of Governmental Owners and Operators of Property. 228
    The Public Duty Doctrine: The Duty to Protect All That Is Enforceable
    by None 229
    Recreational Immunity: The King Can Still Do No Wrong 230
Conclusion 232
Endnotes 232

One evening just after supper in 2002, two mothers in Phoenix, Arizona, lovingly lowered their young sons into warm baths. The mothers were unknown to each other though they lived a mile apart. One mother was Hispanic, the other Anglo. Both were middle class and worked outside the home, like their husbands, to support their families. They let their boys splash happily in their tubs with their water toys, but neither could see the hidden danger in the warm, soapy waters. Some of the bathwater entered the boys' noses. The same domestic water company supplied both homes with the water for those tubs and negligently failed to maintain proper levels of disinfectant in the water. Within three days the boys were complaining of headaches and nausea. Within a week, they were dead.

In 2005, in Tulsa, Oklahoma, two mothers unknown to each other living in the same neighborhood took their children to a splash pad operated by their city in a local park. Both were African American. Both were working class and, like their husbands, worked outside the home to support their families. Both watched over their boys like hawks, but the invisible danger in the water eluded their sharp, young eyes. The children ran through the water, jumping through the sprays with grins on their faces. Some of the water got up their noses. Within days, the boys were complaining of headaches. A few days later, they were both dead.

In 2012, in Stillwater, Minnesota, family members took their child to a swimming facility owned and operated by their city as a park. They were Caucasian and middle class. They were careful and made their boy wear a life vest in compliance with the warning signs at the facility, but no signs warned of the real danger lurking in the invitingly warm waters. They watched over him as best they could as he splashed happily about in the warm, shallow waters by the beach the city had constructed. Within days, the boy complained of feeling sick, and then died.

What no one in authority thought to warn the families about was that two years earlier, in 2010, a young girl also died after swimming in that same pond. Authorities knew why, too. She died from the same type of brain infection caused by the same kind of carnivorous ameba that had killed the two boys in 2002 in Arizona, in 2005 in Oklahoma, and would kill again in 2012 in Minnesota.

What was the cause of these untimely and tragic deaths of young children? The medical cause of death in all five cases was primary meningoencephalitis, PAM for short; a lethal inflammation of the brain due to rapid formation of destroyed brain tissues for which there is no known cure. The fatality rate is close to 99.9 percent. But what was the cause of these five cases of PAM? Thousands upon thousands of carnivorous amebae that invaded the cranial space of the victims and destroyed their brains from inside of their skulls, amebae named *Naegleria fowleri*, smaller than the width of a human hair, but deadly—a virtually invisible killer whose victims number in the hundreds the world over. This chapter is the story of that killer and the laws to which the families of victims must turn for justice.

NAEGLERIA FOWLERI: LEADING CAUSE OF DEATH FROM RECREATIONAL EXPOSURE TO UNTREATED FRESH WATER IN THE UNITED STATES

The leading cause of death from recreational exposure to untreated fresh water in the United States is a little-known single cell ameba.[2] The organism is named *Naegleria fowleri* ("nie-glare-ee-a foul-er-i"). It was first reported upon in the scientific literature in 1965 by M. Fowler and R. F. Carter in Australia as the cause of four deaths, three children ages eight, eight, and nine, and one adult age twenty-eight.[3] The ameba causes PAM, which is virtually 99.9 percent fatal. The United States Centers for Disease Control (CDC) has reported that between 1965 and 2008, *Naegleria fowleri* has taken the lives of more than 133 people, mainly children, in the United States alone.[4] Since 2008, the ameba has reportedly taken the lives of another twenty named victims and five unnamed ones in the United States alone. It has killed even more victims throughout the globe, in Europe and elsewhere.[5]

Naegleria fowleri is a unicellular ameba that is so small that it cannot be detected by the unaided human eye. It is one-third the width of a human hair in size.[6] It prefers soils and warmer waters for habitat, but also endures cold while sheltering in bottom sediments of rivers and lakes. In the wild, it feeds on bacteria or other such micro-organisms. It is not a parasite that lives off of its host, but a free-living organism.[7] It has been found in environmental samples of water from chlorinated swimming pools, untreated freshwater lakes, thermal springs, domestic water supplies, sewage, soil, air, and human nasal passageways. Stormwater runoff is a recognized mode of introducing the amebae to lakes and streams.[8] The ameba is a "shapeshifter," capable of taking any one of three forms of life when environmental conditions warrant.[9]

In its pathogenic trophozoite form (shown in Figure 11.1), the ameba can cause a dangerous PAM, but not when it takes either of its other two forms, a flagellate or a

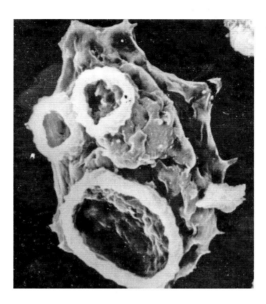

FIGURE 11.1 *Naegleria fowleri.* (Used by permission of Dr. Francine Marciano-Cabral.)

cyst. The route of transmission is usually nasal after which the amebae move to the brain where their feeding activities cause PAM.[10]

PAM is a condition of the brain characterized by the *Naegleria* trophozoites killing brain tissue through a chemical process, called *lysing*, which causes inflammation of the brain that leads quickly to death if not stopped.

The amebae lyse target tissues by excreting active chemicals from round white "sucker cups" that dissolve the cells of the target and allow the ameba to access their contents. The amebic attack has been imaged with electron microscope technology.

Figure 11.2(a) shows neuroblastoma cells cultured in the absence of the ameba trophozoites. After introduction of the amebae, Figure 11.2(b) shows a *Naegleria fowleri* attacking the target cell at the arrow. At eighteen hours post-inoculation, Figure 11.2(c) shows numerous "scars" on the cellular body from attacks by the trophozoites. Six hours later, Figure 11.2(d) shows the remnant of the destroyed target cell.[11]

Legally, the lysing action of the amebae shown in the microimages would appear to satisfy any requirements of law that the amebae present a danger of imminent bodily injury. Once the pathogenic trophozoites enter the nasal passageways, they begin to feed by means of lysing of the mucosal tissues. The lysing actions of the amebae results in damage to cells of the victim and the amebae ingest the contents of the lysed cells as nutrients.

After entering through the nasal passage, the trophozoites cause an infectious process that destroys tissue as the amebae multiply and migrate up the nasal passages, feeding as they go upon the delicate membranes of the nose and sinuses, called *mucosa*. The tissues of the mucosa membranes are also especially delicate and attractive to the amebae as a food source and are located especially close to certain openings in the skull (the *olfactory foramina*), through which pass the olfactory

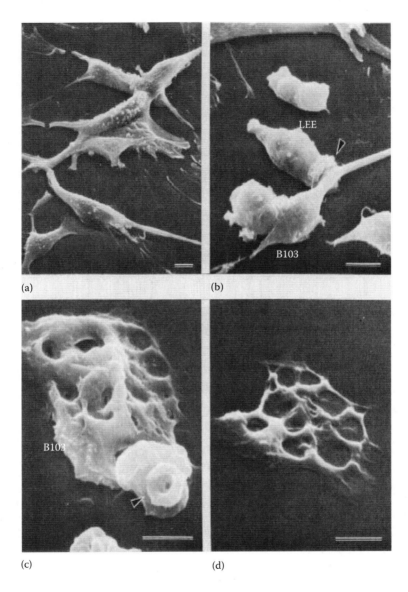

FIGURE 11.2 Scanning electron micrographs of B103 neuroblastoma cells cultured in the absence or presence of *N. fowleri* LEE. (Used by permission of Dr. Francine Marciano-Cabral.)

nerves that connect directly to the brain's tissue, as shown. The mucosa membranes of the nose are also an exceptionally dangerous location for an infection of *Naegleria fowleri* due to the proximity of those tissues to the gateways to the brain and the unusually open access to the brain tissue in that location.[12]

While feeding on the victim's cellular tissues, which they destroy by lysing, the amebae migrate a very short distance (about one inch) along the olfactory nerves and through the openings of the olfactory foramina to gain access to the brain tissue,

Justice for Deaths Caused by Free-Living Amebae in Recreational Waters 227

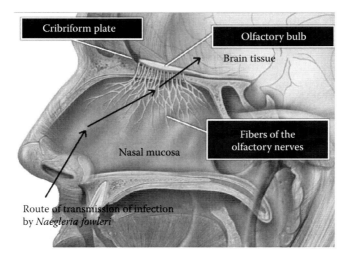

FIGURE 11.3 Route of transmission of infection by *Naegleria fowleri*.

where they then feed on brain tissue. Figure 11.3 accurately shows how close the nasal mucosa are to the openings in the skull (cribriform plate). The brain is exceptionally vulnerable to *Naegleria fowleri* due to the proximity of the nasal mucosa to the olfactory foramina that provide open access to the brain tissue. The ameba's protective behaviors enhance the danger because they cluster together to deny access by the victim's immune response antibodies to the amebae protected in the center of the cluster.

Other than the pathogenic trophozoite form discussed previously, *Naegleria fowleri* can adopt one of two other forms: flagellate or cyst, both of which are nonpathogenic. Only the trophozoite form is flesh-eating and deadly to humans. As flagellate or cyst, the ameba does not feed and is not deadly to humans, but such forms of the ameba can transform back into the pathogenic trophozoite form from the migrating cyst or flagellate forms.

The flagellate form enhances the ameba's locomotion by deploying flagellates from the rear to create motile force, as shown in the right-hand image of Figure 11.4. In the trophozoite form, the ameba moves by extending portions of its membrane

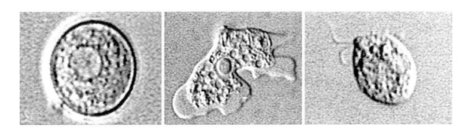

FIGURE 11.4 Three forms of *Naegleria fowleri*. (From Brain-eating amoeba rattles nerves in Louisiana, *Deccan Chron.*, Sept. 20, 2013.)

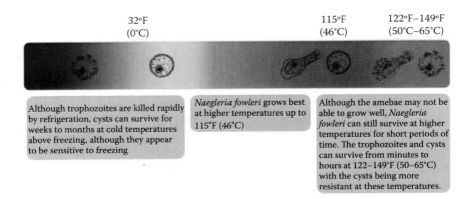

FIGURE 11.5 Environmental conditions for *Naegleria fowleri*. (From U.S. CDC, *Naegleria fowleri*, Pathogen & Enviro., http://www.cdc.gov/parasites/naegleria/pathogen.html.)

(called pseudopods) and then flows into those areas, as shown in the middle image of Figure 11.4. However, when food becomes scarce, the ameba can adopt the flagellate form for enhanced motion and can switch back when conditions become more favorable to feeding.[13]

The nonpathogenic cyst form shown in the leftmost image of Figure 11.4 enables the amebae to endure harsh environmental conditions including food scarcity and temperature stressors. The cysts can transform back into trophozoites or flagellates under improved conditions.[14] The cysts can be carried in dusts.[15] There is one reported case of a PAM infection from airborne cysts.[16]

The cyst form of *Naegleria fowleri* is more resistant to cold than the trophozoite form. Cysts have been reported to survive cold temperatures down to 0°C (absent freezing) for months, but survival time drops precipitously to under an hour when temperatures plunge below freezing.[17] Thus, cysts may be able to survive winter conditions by sheltering in the sediments in deepwater lakes where the lakes do not freeze. See Figure 11.5.

Naegleria fowleri can be killed by chlorination levels and contact times that are well within the practical reach of disinfection systems. Levels of chlorination between 0.5 and 1.0 ppm with contact times of 1 hour have been shown effective,[18] though of course chlorine degrades over time and under sunlight and other environmental factors.

DUTIES OF GOVERNMENTAL OWNERS AND OPERATORS OF PROPERTY

To illustrate the legal principles applicable to cases in which families and next of kin seek justice for victims of PAM by *Naegleria fowleri*, this chapter will use the Minnesota case described in the introduction, which involves a municipal owner/operator of land that supplies water for recreational use.[19] In many states, political subdivisions like cities and towns enjoy a more protective body of law in this regard than do private landowners.

THE PUBLIC DUTY DOCTRINE: THE DUTY TO PROTECT ALL THAT IS ENFORCEABLE BY NONE

The majority of states adhere to the so-called "public duty doctrine." It provides that duties that a state or political subdivision owes to all members of the public at large are not legally enforceable by any injured member of that public.[20] The public duty doctrine is basically the old doctrine of sovereign immunity ("the king can do no wrong") under an assumed name because the results are the same: The state or political subdivision that has caused injury or death cannot be held liable for its misconduct. The injustice of such a result is manifest from the fact that the public entities are able to spread the risks among their public beneficiaries of their services by obtaining the same kind of liability insurance that private operators of water parks can obtain. Rather than shift the losses of its operations to the beneficiaries of parks and public water facilities by the vehicle of insurance, the public duty doctrine imposes the full burden of such tragic losses on those least able to bear them, the bereaved families of the slain. A number of states have abrogated, limited, or abolished the doctrine.[21]

Noxious as it is in its results to injured persons seeking justice, the public duty doctrine is subject to some limits that advocates for states, cities, and towns may be tempted from time to time to forget or conceal. The main limitation is that the doctrine is (and should be) limited to the fact patterns of the cases from which it arises: Cases involving harm to an injured party from the actions of some third person that the public entity was in a position to prevent but did not do so. A typical example arises in the building code cases in which a person injured by substandard construction by the primary defendant seeks to hold the city building department responsible for reviewing obviously substandard plans. These cases generally exonerate the building department from any duty to protect homeowners from the negligence of building contractors unrelated to the cities, but these holdings should be limited to their facts. The Supreme Court of Montana in *Gaitlin-Johnson v. City of Miles City*, 367 Mont. 414, 291 P.3d 1129 (MT 2012) best described the limitation as

> The public duty doctrine was **not intended to apply in every case to the exclusion of any other duty a public entity may have**. It applies only if the public entity truly has a duty owed **only** to the public at large…. **It does not apply where the government's duty is defined by other generally applicable principles of law**. (291 P.3d at 1133 [public duty doctrine inapplicable where independent source of duty from premises liability law; emphasis added].)

As the Minnesota Supreme Court put it in *Cracraft v. City of St. Louis Park*, 279 N.W.2d 801, 803 (Minn. 1979), "We are, instead, considering the municipality's ***unique*** duty to enforce the law ***by taking steps to assure that third persons comply with the law***" (emphasis added). Thus, the doctrine distinguishes between a specific duty to warn about or remove risks created or maintained ***by others*** from risks of harm created or maintained ***by the public agency itself***.[22] *Cracraft v. City of St. Louis Park*, 279 N.W.2d 801 (Minn. 1979), cautioned that the public duty doctrine did not apply to governmental agencies that control or own public facilities because *these agencies already owe common law duties of care to individuals by virtue of that control*: "We are not concerned with the legal duties owed by municipalities as owners and operators of buildings, roadways, or other facilities" (279 N.W.2d at 803).

Injured persons seeking recovery for the deaths of loved ones from PAM due to *Naegleria fowleri* in water supplied by public agencies should be aware of the pitfall posed by the public duty doctrine and couch their claims accordingly. They should focus on negligence on the part of the public agency itself instead of any failures by public agencies to prevent harms caused by third persons.

Recreational Immunity: The King Can Still Do No Wrong

As sovereign, King Arthur enjoyed immunity from lawsuits by farmers whose crops were damaged by the Round Table's knights riding roughshod wherever they liked in pursuit of the Holy Grail. Twelve centuries later, the legal doctrine of sovereign immunity is just as unjust and unfair as it was back then.[23]

Most states in the United States have waived sovereign immunity by statute, but have allowed it to creep back in through exceptions to those waivers. One of the most prevalent exceptions is "recreational user immunity," that is, immunity for park operations. Minnesota's statute is typical in that it excepts recreational users of parks from the waiver of sovereign immunity (thereby restoring the immunity), but creates an "exception to the exception" so that public agencies owe the same legal duties as private landowners owe to trespassers.[24]

When contesting recreational immunity, litigants must bear in mind the applicable rules of construction for exceptions to exceptions as the proper application of the rules of construction can be the decisive difference in victory and defeat. The exception to the waiver of sovereign immunity is typically required to be construed broadly in accordance with the remedial purposes of the waiver. Thus, recreational immunity must be construed narrowly because it is an exception to the exception that reinstates the rule of immunity. Again, Minnesota law is typical.[25]

The exception to recreational immunity for trespassers is generally construed in accordance with the *Restatement of Torts*, Sections 335 (adults) and 339 (unaccompanied minors). As is typical among the states adopting recreational immunity, Minnesota provides that public park operators owe the invited users a duty to exercise a minimal level of care for their safety—the "same duty that a private person owes to trespassers …," that is, to warn of hidden, or concealed, dangers, of which the public agency was, or should have been, aware.[26]

Section 335 of the *Restatement* states the rule of trespasser liability for adults and children supervised by adults. A possessor of land will be liable for serious bodily harm caused by (1) "an artificial condition of the land" known to the possessor if he or she (2) creates or maintains that condition, (3) has reason to believe that trespassers would not discover it, and has (4) negligently failed to warn of the risks.

Section 339 of the *Restatement* states the rule of trespasser liability for unsupervised children. A possessor of land will be liable for physical harm to children caused by an "artificial condition upon the land" if (1) he or she knows or should know of it, (2) it is unreasonably dangerous to children who because of their youth would not discover it, (3) the utility of maintaining it and the burden of eliminating it are slight compared to the risk, and (4) the possessor fails to exercise reasonable care to eliminate it.

Litigants seeking to capture the benefits of the "trespasser exception" should keep in mind the doctrine of proximate cause as their opponent will probably advocate the most limited interpretation of what constitutes an "artificial condition" that caused the injury. For example, if a city has constructed improvements to a pond for public swimming, the city may be subject to the trespasser exception once the doctrine of proximate cause is taken into account.

In the Minnesota case on which the example in this chapter's introduction is based, the city constructed certain improvements to the pond to facilitate swimming that also facilitated the population of *Naegleria fowleri* in the pond that had killed the young girl in 2010:

1. A limited area of beach set aside by the city for swimming that also created a shallow zone of hot water in the summer months that was unusually conducive to the transformation of non-pathogenic *Naegleria* cysts hiding in the bottom sediments into waterborne, pathogenic, *Naegleria* trophozoites, right in the area limited to swimming.
2. Storm water piping that introduced the *Naegleria* into Lily Lake and conveyed unusually high levels of nutrients in the untreated, urbanized run-off into the lake so as to artificially stimulate the amebae cysts in the beach sediments to "hatch out" into the pathogenic trophozoite form of life right in the area limited to swimming.
3. Thirteen warning signs at the pond addressing the risks of drowning, animals, and milfoil weed, but omitting to mention the risk of death from *Naegleria* and thereby raising the negative inference in the public's mind that all substantial dangers of death had been disclosed and the lake was otherwise safe for swimming, though it was not.

In rebuttal to the contention that no "artificial condition" caused the boy's death (but rather the amebae was responsible), the doctrine of proximate cause was implicated. This doctrine defines as a legal cause of an injury any condition or configuration that is a "substantial factor" even though it is not the only factor.[27] In addition, proximate causation is a question of fact for the jury.[28]

Substantial factor, proximate causation is also embedded in Sections 335 and 339 of the *Restatement*, both of which require causation for liability.[29] In addition, Section 430 of the *Restatement* states, "In order that a negligent actor shall be liable for another's harm, it is necessary ... that the negligence of the actor be a legal cause of the other's harm." The *Restatement* defines "legal cause" as "The actor's negligent conduct is a legal cause of harm to another if ... his conduct is a substantial factor in bringing about the harm ...." (Section 431, *Restatement*).

In the example case from Minnesota, the beach area, ramadas, BBQs, and other improvements at the park together with the warning signs constituted a "substantial factor" in the exposure to and subsequent death from *Naegleria fowleri*. Therefore, they constituted an "artificial condition" on land under the exception for trespasser liability.

One of the typical counterarguments by public agencies under these circumstances is to raise the old specter that broad interference with government services will result from allowing families of injured children to pursue justice. The Minnesota Supreme Court has provided an instructive rebuttal:

> [I]n no instance, even where immunity is not recognized, has a municipality been seriously handicapped by tort liability. This argument is like so many of the horribles paraded in [20] the early tort cases.[30]

In this age of sophisticated risk pooling and reinsurance mechanisms to spread risk at affordable premiums to the public at large that enjoys the recreational benefits of parks, such an argument is a mirage intentioned to deter courts from doing justice. It is only fair that the general public which enjoys general access to recreational amenities such as parks should bear the costs of negligent operation by their owners, whether public or not, instead of allowing such tragic loss to fall solely on an individual family.

CONCLUSION

The risks posed by *Naegleria fowleri* to recreational users of fresh water facilities offered as parks by private and public operators are not well known throughout the country. Though small in absolute number of fatalities, the impact on the victims and their families far outweighs the numbers of cases. The applicable legal doctrines have not caught up with the advancing state of scientific knowledge about this invisible killer.

ENDNOTES

1. The author is a lawyer practicing in Arizona and Nevada. He represented the families of Zachary Stalls and Davy Luna, of Phoenix, Arizona, killed in 2002 by *Naegleria fowleri* in a domestic water system; the families of Martinez Owens and Terrell of Tulsa, Oklahoma, killed in 2005 by *Naegleria fowleri* in the water system of a municipal splash pad; and the family of Jack Ariola, killed in 2012 by *Naegleria fowleri* in a municipally operated swimming facility in Stillwater, Minnesota. The Ariola matter is the subject of a state court appellate decision: *Ariola v. City of Stillwater*, 2014 Minn. App. Unpub. LEXIS 1136 (Minn. Ct. App. Oct. 27, 2014). Consulting on all of these cases was Dr. Francine Marciano-Cabral of the Medical College of Virginia at Virginia Commonwealth University in Richmond, Virginia. She is one of the world's authorities on *Naegleria fowleri*.
2. In U.S. English, according to the CDC, the single-celled living organism described here is an *"ameba."* The word *"amoeba,"* with an "o", is used as part of a scientific genus name (such as *Amoeba* or *Acanthamoeba*). In British English, both the generic organism term and genera names are spelled *"amoeba"* with an "o".
3. Fowler, M. et al. 1965. Acute pyrogenic meningitis probably due to Acanthamoeba sp.: a preliminary report. *Br. Med. J.* 2:740–2.
4. Yoder, J.S. et al. 2010. The epidemiology of primary amoebic meningoencephalitis in the USA, 1962–2008. *Epidemiol. Infect.* 138:968–75. Other reported fatalities in years since 2008 include the following: (2009) **Dalton Nettles**, age 9, after swimming

in a lake in Madison County, FL; **Philip Thomas Gompf**, age 10, possibly after swimming in Polk County's Lake Arietta, FL; **Shane Sugaski**, 22, of Lake Wales, FL; (2010) **Davian Briggs**, age 7, of Little Rock, AK; **Lisa Hollingsworth**, age 10, of Mt. Pleasant, SC; **Will Matthews**, age 14, allegedly after wakeboarding in Cross Lake, LA; (2011) **Courtney Nash**, age 16, after swimming in the St. John River near Orlando, FL; **Andrew Chavez**, age 14, after swimming in Winfield City Lake, KS; **Jeffrey Allen Cusimano**, age 28, of Arabi, LA, after alleged exposure to domestic water while using a "NasaFlo Neti Pot" to wash his nasal cavities; **Mason Fauble**, age 6, of Missouri Valley, IA, after swimming in a MN lake; **Annie Bahneman**, age 7, from swimming in Lily Lake in Stillwater, MN; **Kyle Gracin Lewis**, age 7, after swimming in the Paluxy River at Dinosaur Valley State Park in Glen Rose, TX; **Christian Alexander Strickland**, age 9, of Henrico County, VA, after swimming in the James River; (2012) **Blake Diggers**, age 8, after swimming in Lake Marion, SC; **Waylon Abel**, age 30, allegedly after swimming in West Boggs Lake, IN; an **unnamed Bryan Co., OK, youngster** after swimming in the Red River; **Jack Ariola**, age 9, from swimming in Lily Lake in Stillwater, MN; (2013) **Zachary Reyna**, age 12, after waterboarding in a ditch by his house in FL; **Drake Smith, Jr.**, age 4, after playing in a lawn water slide supplied by domestic water in LA; (2014) **Hally Yust**, age 9, after swimming in several lakes in KS; (2015) **Kelsey McClain**, age 24, after swimming in the Colorado River near CA; an **unnamed woman, age 21**, from CA who passed away on June 20, 2015, at Renown Medical Center in Reno, NV; an **unnamed Oklahoma City woman** after swimming in Lake Murray in OK; an **unnamed Oklahoma man, age 31**, after swimming in Lake Murray, OK; **Michael John Riley Jr.**, age 14, after swimming at Sam Houston State Park, in TX. In addition, a fatality in 2008 involving **Hailee LaMeyer**, then 11, after swimming in a lake in MN has also been connected by the family to *Naegleria fowleri*. In addition, in 2013, **Kali Hardig**, age 12, was exposed allegedly from swimming at Willow Springs Water Park in south Little Rock, AK, and survived, one of three reported survivors.

5. E.g., Ahmed, T., *Naegleria* Claims Its Eleventh Victim in Sindh [Province of Pakistan]. *DAWN NEWS*, July 28, 2015, http://www.dawn.com/news/1196905; Cogo et al., Fatal *Naegleria fowleri* Meningoencephalitis, Italy, *Emerg Infect Dis.* 2004 Oct;10(10):1835–1837 [1st case in Italy, 9-year-old boy after swimming in Po River in Northern Italy].
6. The size of a typical *Naegleria fowleri* trophozoite has been reported as 10–15 μm in diameter, (see Martinez, A. J. 1996, Free-living amebas: *Naegleria, acanthamoeba* and *balamuthia, Medical Microbiology*, 4th ed., Galveston, TX: The University of Texas Medical Branch at Galveston [Hereinafter, Martinez (1996)]), which is roughly a third of the diameter of a typical human hair.
7. For an in depth treatment, see, Marciano-Cabral, F., 1998, Biology of *Naegleria fowleri* spp., 52 *Microbio. Rev.*, 114–133 [hereinafter, "Cabral (1988)"].
8. Cabral (1988) at p. 115.
9. Cabral (1988) at p. 115.
10. Cabral (1988) at p. 115.
11. Note, Marciano-Cabral, F. et al., 1983, Cytopathogenicity of *Naegleria fowleri* for rat neuroblastoma cell cultures: Scanning electron microscope study, 40 *Infect. and Immun.*, 1214–17.
12. Cabral (1988) at p. 116.
13. Visvesvara, G., Yoder, J., Beach, M.J. 2012. Primary amebic meningoencephalitis, in *Netter's Infectious Diseases*, eds. Yong, E.C., Stevens, D.L., pp. 442–447. Philadelphia, PA: Elsevier Saunders.
14. Visvesvara, G.S., Moura, H., Schuster, F.L. 2007. Pathogenic and opportunistic free-living amoebae: *Acanthamoeba* spp., *Balamuthia mandrillaris, Naegleria fowleri*, and *Sappinia diploidea*. 50 FEMS *Immunol Med Microbiol.* 1.

15. Martinez (1996).
16. Cabral (1988) at p. 116.
17. Chang, S.L. 1978. Resistance of pathogenic *Naegleria* to some common physical and chemical agents. 35 *Appl Environ Microbiol.*, 368–75.
18. De Jonckheere, J., van de Voorde, H. 1976. Differences in destruction of cysts of pathogenic and nonpathogenic *Naegleria* and *Acanthamoeba* by chlorine. 31 *Appl Environ. Microbiol.*, 294–7 (1mg/l for 1 hour for cysts).
19. Private landowners who own or operate recreational facilities that supply water for the use of the public such as water parks generally owe duties to their business invitees to take reasonable protective measures against latent hazards that the owner/operator knows or should know are present on the property either by posting appropriate warnings, or, in the case of children, acting reasonably to remove such hazards. For example, the *Restatement (Second) of Torts* § 343 provides that a possessor of land is liable for physical harm caused to his invitee by a condition on land if, but only if, he or she (a) knows or by the exercise of reasonable care would discover the condition, and should realize it involves unreasonable risk of harm to such invitee; (b) should expect that they will not discover or realize the danger, or fail to protect themselves against it; and (c) fails to exercise reasonable care to protect them against the danger.
20. The public duty doctrine is the law in a majority of states: **Maryland** (*Muthukumarana v. Montgomery Cty.*, 370 Md. 447, 805 A.2d 372 [2002; local government emergency telephone system employees did not owe an individual tort duty to persons in need of their services; therefore, they could not be liable for the negligent performance of a duty]); **Michigan** (*White v. Beasley*, 453 Mich. 308, 552 N.W.2d 1 [1996; adoption of the public duty doctrine], *Beaudrie v. Henderson*, 465 Mich. 124, 631 N.W.2d 308 [2001; Public duty doctrine was not extended to protect governmental employees, other than police officers, alleged to have failed to provide protection from the criminal acts of third parties, especially given the governmental immunity statute]); **Missouri** (*Southers v. City of Farmington*, 263 S.W.3d 603 [Mo. 2008]); **Montana** (*Kent v. City of Columbia Falls*, 379 Mont. 190, 350 P.3d 9 [2015; despite applicability of public duty doctrine, the city had a duty from statute and voluntary undertaking]); **Rhode Island** (*Torres v. Damicis*, 853 A.2d 1233 [R.I. 2004; When worker was hurt while working on a project for which a building permit was erroneously issued, the public duty doctrine immunized town whose building inspector issued permit, and neither the special duty nor egregious conduct exceptions applied]); **North Carolina** (*Multiple Claimants v. N.C. HHS, Div. of Facility Servs.*, 361 N.C. 372, 646 S.E.2d 356 [2007; the special relationship exception to the public duty doctrine applied to the inmates' negligence claims arising from a fire in a county jail where the relevant statutes and regulations showed that the Department of Health and Human Services had a duty to inspect local jails to ensure that they met the minimum fire safety standards]); **Tennessee** (*Ezell v. Cockrell*, 902 S.W.2d 394 [Tenn. 1995; a police officer allowed a man to drive, who then hit the victim's truck, hurt her, and killed her husband. Under the public duty doctrine, the officer was shielded from liability to the victim because his duty was not to her only but to the public]); **Utah** (*Cope v. Utah Valley State Coll.*, 2014 UT 53, 342 P.3d 243 [2014; declining to abrogate public duty doctrine]); **Washington** (*Munich v. Skagit Emergency Commc'ns Ctr.*, 175 Wn. 2d 871, 288 P.3d 328 [2012; where a decedent's estate alleged a 911 operator negligently responded to an emergency call, express assurances promising action need not be false or inaccurate as a matter of law to satisfy the special relationship exception to the public duty doctrine]); **West Virginia** (*Holsten v. Massey*, 200 W. Va. 775, 490 S.E.2d 864 [1997]).
21. **Colorado** (*Leake v. Cain*, 720 P.2d 152 [Colo. 1986; in reviewing a drunk driving wrongful death case, the court abolished the public duty doctrine]); **New Hampshire** *Doucette v. Town of Bristol*, 138 N.H. 205, 635 A.2d 1387, 1390 (N.H. 1993; "[T]he public duty rule

impermissibly conflicts with the abrogation of common law municipal immunity...."); **Alaska** (*Adams v. State*, 555 P.2d 235, 241 [Alaska 1976; "[W]e consider that the [public duty] doctrine is in reality a form of sovereign immunity, which is a matter dealt with by statute in Alaska, and not to be amplified by court-created doctrine"]); **Ohio** (*Wallace v. Ohio DOC*, 2002-Ohio-4210, 96 Ohio St. 3d 266, 773 N.E.2d 1018 [2002; in suit against fire marshal for not inspecting fireworks store, public-duty rule was incompatible with statute requiring state's liability in court of claims to be determined in accordance with rules applicable to suits between private parties]); **Wyoming** (*Natrona Cty. v. Blake*, 2003 WY 170, 81 P.3d 948 [Wyo. 2003; in a wrongful death action alleging that the county owed a duty to a deceased, who was killed by a jail inmate allegedly allowed to negligently escape, a duty on the part of the custodians did exist]).

22. Compare: *Cracraft v. City of St. Louis Park*, 279 N.W.2d 801 (Minn. 1979; failure to discover during inspection certain violations of city fire code by someone else not actionable absent a special relationship); to *Domagala v. Rolland*, 805 N.W.2d 14, 24 (Minn. 2011; "general negligence law imposes a general duty of reasonable care when the defendant's ***own conduct*** creates a foreseeable risk of injury to a foreseeable plaintiff.")

23. Professor Erwin Chemerinsky, from whom countless thousands of lawyers have learned constitutional law in review courses for bar examinations across the country, has characterized sovereign immunity as an "anachronistic relic" that should be entirely eliminated from American law. Chemerinsky, E. 2001. Against sovereign immunity. 53 *Stanford L. Rev.* 1201–1224.

24. Generally, states and cities are liable for their torts (Minn. Stat. §466.02). Exceptions are provided (Minn. Stat. 466.03). The recreational immunity is provided in subjection 6e which immunizes municipalities from "Any claim based upon the construction, operation, or maintenance of any property owned or leased by the municipality that is intended or permitted to be used as a park, [or] as an open area for recreational purposes of for the provision of recreational services, ..." The exception to the exception is set forth in the same subjection as "Nothing in this subdivision limits of liability of a municipality for conduct that would entitle a trespasser to damages against a private person...."

25. E.g., Minn. Stat. 466.02, *Doyle v. City of Roseville*, 507 N.W.2d 33, 35 (Minn. App. 1993; since the court must "narrowly construe exceptions to the general rule of municipal tort liability," the terms "park," "open area," or "recreational services," in the recreational immunity exception contained in Minn. Stat. 466.03, subd. 6e, did not include parking lot adjacent to ice arena, city not immune), *reversed on other grounds*, 524 N.W.2d 461 (Minn. 1994). (App. 099).

26. See e.g., *Lundstrom v. City of Apple Valley*, 587 N.W.2d 517, 519-20 (Minn. App. 1998; tennis player slipped due to duct tape in plain view that he had previously observed on court surface, city immune); Minn. Stat. §466.03, subd. 6e, and Sections 335 and 339 of the *Restatement (Second) of Torts* ("*Restatement*").

27. E.g., *George v. Estate of Baker*, 724 N.W.2d 1, 10 (Minn. 2006; "Minnesota applies the substantial factor test for causation. The negligent act is a direct, or proximate cause of harm if the act was a substantial factor in the harm's occurrence.").

28. E.g., *Osborne v. Twin Town Bowl, Inc.*, 749 N.W.2d at 373 ("Whether proximate cause exists in a particular case is a question of fact for the jury to decide.").

29. Section 335: "A possessor of land ... is subject to liability for bodily harm ***caused to*** [trespassers] by an artificial condition...." Section 339: "A possessor of land ... is subject to liability for physical harm to children trespassing thereon ***caused by*** an artificial condition ...."

30. *Spanel v. Mounds View Sch. Dist.*, 264 Minn. 279, 288–89 (1962; quoting the Illinois Supreme Court).

12 The Psychological Evaluation
Trauma Caused by a Serious Aquatic Injury or Fatal Drowning

Diana P. Sullivan

CONTENTS

Treatments 244
    Cognitive Behavioral Therapy (CBT) 244
    Neuropsychological Evaluation 245
Relaxation Therapy 245
Eye Movement Integration (EMI) Therapy and Eye Movement Desensitization and Reprocessing (EMDR) 246
Medication 246
Physical Therapy 247
Attendance of Support Groups 247
References 247
Websites 248

Most serious injuries and accidental deaths have a severe psychological impact on the afflicted individual and on their family members. Mental trauma or psychological injury has been underdiagnosed as a basis for litigation, and attorneys often overlook the need to screen accident victims and family members. Aquatic injuries and fatal drownings call for significant attention to this issue, due to the fact that most aquatic incidents and mortalities are often preventable. Awareness of preventability of accidents can intensify grief of those experiencing loss and significantly prolong and deepen the healing process.

    The most common mental health issues associated with loss or injury include, but are not limited to, anxiety disorders, depressive disorders, trauma and stressor-related disorders, major or mild cognitive disorders due to a traumatic brain injury, and substance-related and addictive disorders. Criteria for diagnosing a specific disorder have been established by the American Psychiatry Association. *The Diagnostic and Statistical Manual of Mental Disorders*, fifth edition (*DSM-V*), is used by most mental health care professionals for diagnoses.

Grief is a universal human experience, a recurrent theme as we experience life; however, not all losses have an emotional impact on the individuals. Grief is defined as "a profound feeling of sadness and pain caused by an important loss, change, crisis, or failure either actual or perceived" (Holland, 2011). Some life changes and major losses commonly lead to experiencing grief. These changes may include the loss of a relative or friend, disability, permanent injury or terminal illness, aging, job loss or retirement, divorce, financial change, jail term, and so on.

Grief is differentiated from other losses because it includes emotional pain, therefore the love and meaning associated with that particular loss plays an important role in the intensity of the grief experienced. The deeper and the more significant the love or attachment is, the more complicated the grief process. No one is exempt of experiencing pain or suffering through losing someone such as a relative or friend.

Historically, the grief process encompasses five different stages for the individual to come to grips with his or her condition. This concept has its validity when dealing with a person who has been diagnosed with a terminal illness; however, it does not apply in its totality for a person who has suffered a traumatic event or when bereavement is present. These stages of grief will be discussed further in this chapter.

Clinical data reveal that it is not conclusive to extrapolate findings related to grief to a particular drowning event. Grief is emotional, is often an unknown reaction, and it does not always make sense. The symptomatology of grief is unique to the individual and no two outcomes are the same. These reactions will affect the individual major life's spheres and not necessarily all the symptomology are inclusive or exclusive to all grievers. Grief usually catches people off guard; due to its unexpected nature it is an inimitable experience, and no two events are the same or are the people who experience them.

Each situation has its own unique manifestations with multiple variables that must be considered before an evaluation can be objectively reached with scientific confidence. The main point or crucial factor is that a serious injury or fatal drowning must be evaluated separately and objectively without any preconceived beliefs or bias about clinical outcomes.

This has particular significance in cases where attorneys are alleging negligence. Juries must have valid and reliable information to deliberate verdicts and to determine the extent of economic damages. Only a clinical evaluation will reveal this information among the severely injured with long-term medical and psychological interventions. Clinical assessments and treatment may be ongoing, making short-term evaluations problematic. This is of the utmost importance among victims suffering a spinal cord injury where there is no hope of recovery and the realization of an abbreviated lifespan. Also, just as important is the devastating impact it will have among family members.

Another term important to address is trauma. *Trauma* is defined as an outside event out of the normal range of our human experience. It is an event that occurs without any warning. It is often associated with a state of shock, disbelief, and surrealism; and is accompanied by intense fear, terror, and helplessness. If the trauma is of a personal nature or against an immediate family member, the symptoms are more intense and long-lasting. Trauma symptoms can last for a few days to a lifetime. If the symptoms are less than month's duration, the diagnosis of acute stress disorder is warranted. However, if the symptoms continue for a longer period of time, the diagnosis of posttraumatic stress disorder (PTSD) will be given. It includes dissociative symptoms such as depersonalization and derealization, according to the *DSM-V*.

It is important to notice that the onset of full-spectrum PTSD can be delayed for months or even years; however, some symptoms will be evident within a week to three months following the traumatic event.

Either grief or a traumatic event may cause an individual significant distress that will affect physical, emotional, psychological, social, vocational, and financial areas. An individual who has experienced a loss cannot envision what their grief will be like or how long the recovery process will be.

Grief will range from mild sadness to a continuous reminisce of the incidents that can last for many years. If not treated appropriately it may lead to other more complicated mental health issues such as chronic stress, anxiety, phobias, depression, drug and other addictions, physical illnesses, and even suicide, in extreme cases.

Psychological or psychosocial evaluations must be thorough and gather information in the following areas to determine all contributing factors to the grieving process.

- Age, gender, and culture
- Developmental stage and cognitive abilities
- Personality and attitude
- Religion or spiritual beliefs
- Type and circumstances surrounding the loss
- Family pattern of grieving
- Past responses to previous losses
- Significance assigned to the loss
- Social support
- Hobbies and significant activities in one's life

As an individual experiences grief, dwelling on what has been lost is part of the normal process. The individual is faced with their loss on a daily basis and for some, moving through this process may be very difficult. Grief is an unescapable part of dealing with physical loss.

Physical loss symptomatology has been studied for decades. Experiencing a chronic condition, such as a terminal illness; or becoming handicapped, such as suffering from paraplegia, quadriplegia, a brain injury, or losing an important body function, can disrupt life in many ways. As a result of these injuries, the uncertainty about the future may be devastating. It not only brings emotional, mental, and physical pain, but also the uncertainty about economic stability. The stress that these types of losses bring is ongoing and therefore warrants the assistance or help of many professionals.

When a tragedy strikes it is extremely important for the individual suffering the trauma and their families to consider taking a multidisciplinary approach to their suffering to ensure their well-being and that of their families. It includes medical, psychological, and legal approaches, if appropriate.

Previously the importance of a thorough psychological evaluation was mentioned. Clinical manifestations will be exhibited by someone suffering a chronic condition or when someone suffers dismemberment, their life will be disrupted in many ways and their life style will be altered. Life's simple routines or tasks may be hard to perform and the injured may need the assistance of other family members, a care giver professional, or an assisted living facility.

The physical discomfort experienced by victims after traumatic injuries will have an impact in their emotional state. It is possible that an individual might experience a high level of stress and could experience an overwhelming range of emotions from anger, resentment, guilt, anxiety, sadness, to fear or depression. All these factors are emotionally draining and psychologically debilitating if not addressed or process appropriately.

For some, moving through this process is very difficult. Without professional help or the support of friends and family, the grief process is much more complicated and it is hard to move on.

The following are some of the symptoms experienced by a person dealing with a chronic condition such an illness, the result of an accident, or during the grieving process:

- Reduced physiological activity
- Limited ability or mobility
- Changeable levels of pain and fatigue
- Reduced physical strength and stamina
- Quality of the health insurance
- Limited health insurance resources
- Medication side effects
- Changes in daily routines
- Reduced amount of work or resignation from work
- Financial difficulties or burden
- Anxiety, depression, chemical dependency, or other mental health conditions

There are also secondary losses as a result of a medical condition or permanent body injury. The following have been reported:

- Loss of enjoyable activities from the past
- Loss of independence, sense of security
- Low self-esteem, questions self-worth
- Worries about medical bills and needed devices not covered by insurance
- Questions role in relationships (family, work, community)
- Loss of dreams and hopes for the future
- Isolation in order to protect self
- Guilt, helplessness

Physical effects include the following:

- Change in sleeping patterns
- Gastrointestinal problems

- Change in sex drive
- Change in weight
- Isolation
- Being easily fatigued
- Muscle tightness

Mental effects include the following:

- Memory impairments
- Poor concentration and comprehension
- Struggles with decision making
- Loss of focus
- Vivid dreams and nightmares
- Repetitive, distressing thoughts about the event
- Flashbacks

The individual will experience a wide range of emotions and mood swings. Emotional effects include the following:

- Sadness
- Anger
- Irritability
- Bitterness
- Resentment
- Frustration
- Guilt (real or imaginary)
- Regret
- Shame
- Powerlessness
- Helplessness
- Anguish
- Despair
- Apathy
- Emptiness
- Numbness
- Detachment
- Loneliness
- Fear of going crazy or out of control
- Fear of rejection and abandonment
- Fear of not being able to cope
- Depression
- Anxiety
- Chemical dependency

Regarding drownings in particular and taking into consideration that most drownings are preventable, the feeling of guilt needs to be addressed separately.

Guilt is a feeling of blame and regret that is usually hard to acknowledge and often difficult to express. The thought process of the person experiencing guilt includes the "what if" and "if only" scenarios. These thoughts can be paralyzing and debilitating while the person tries to determine what they could have done differently or how could they "right" a wrong doing.

Behavioral effects include the following:

- Erratic, irrational behavior
- Avoidance, shutting down
- Immersion in work
- Apathy
- Disengagement from previous pleasurable activities
- Excessive drinking
- Difficulties expressing and sharing feelings
- Imagining the deceased in a crowd

Spiritual effects are as follows. Both ends of the spectrum are found in this category:

- Losing faith in God
- Anger toward God and the religious institution
- Feeling that life is not fair
- Feeling that God has betrayed them
- Thoughts of an afterlife
- Strengthening of the faith
- Identification toward others in the same faith
- A new appreciation for life and goodness of people

Bereavement is defined as the state of sadness due to a family member or friend's death; to be deprived by death.

A review of the life expectancy through the centuries indicates that during the Roman era life expectancy was 22 to 25 years. At the beginning of the nineteenth century, life expectancy was 30 years of age. Child birth was the most significant cause for women to have shorter life spans. The cause was a lack of contraceptive devices; therefore, multiple pregnancies were involved. It was only with the development of penicillin and other sulfas in the 1930s that life expectancy significantly increased, to 58 years for men and 62 years for women. In 2006, life expectancy had risen to 75 years; in 2012, to 78.74 years. The current life expectancy for men is 78.8; for women, 81.2 years.

The Kübler-Ross model was the first study that addressed some of the stages that a person goes through when given a terminal illness diagnosis. In her first findings, Elizabeth Kübler-Ross reported that a person goes through progressive stages, in order as follows: Denial, anger, depression, bargaining, and finally acceptance of their condition. Years later Kübler-Ross discovered that there was not a sequential order to the stages but that the individual oscillated among them. The model explains and

The Psychological Evaluation

works well with this population. Although a pioneer is this area, her model has been revised and does not always work well or fit the complex reactions for those grieving a death or other losses. *Grief is an individual experience and no two people grieve alike.*

The events surrounding a death include but are not limited to the type of death, that is, accidental versus natural; the closeness to the deceased; religious/spiritual beliefs; and the cultural view on death. Each of these factors will make the grieving process more or less intense and more or less complicated. Each person is unique, therefore so is their length and depth of bereavement.

Cross-cultural considerations must be taken into account when conducting psychological evaluations. There are significant differences in regards to how grief and bereavement is viewed and expressed cross-culturally. Dr. John R. Fletemeyer estimates that between 6 and 10 percent of aquatic injuries and fatal drowning cases involved victims and families that were either first-generation immigrants or visitors from foreign countries without a Christian background.

Many of these individuals have a religious system that is polytheistic and some are convinced atheists with no religious system (see Figure 12.1). A Gallup International poll involving 65 countries reported that 11 percent of the population does not believe in God whereas a PEW poll elevated that number to 16 percent.

The length and period of bereavement ranges in a wide spectrum. It can end in recovery (most likely) with no improvement or further deterioration. The diagnosis of adjustment disorder is given when the individual will experience a readjustment period of several months to a couple of years. This is normal. Sometimes the bereaved can show no signs of improvement and continue to have symptoms of depression and anxiety; less frequently, the bereaved can deteriorate and experience significant or major psychiatric diagnosis.

FIGURE 12.1 Photo of a religious memorial constructed by family members who had a son or daughter who fatally drowned.

TREATMENTS

Various approaches have been proven to have significant impact in the treatment of these disorders. A multidisciplinary approach involving a team of professionals will make the recovery process much easier and with more effective results. However, it could be costly and time-consuming. This is an important factor to consider when addressing the cost of psychological evaluations and interventions.

The need for ongoing evaluations and treatment is usually recommended. The length of treatment varies from individual to individual and its duration might range from a few months to years before the individual fully recovers. Treatments must address the physical, emotional, behavioral, interpersonal, existential, and spiritual levels. The following are some of the most conventional and successful strategies and techniques used.

COGNITIVE BEHAVIORAL THERAPY (CBT)

The focus of cognitive therapy is to help people identify their distressing thoughts and evaluate how realistic their thoughts are based on the premise that our emotional feelings are based on how we interpret or perceive the situations we experience. In other words, CBT teaches the individual how to identify irrational thoughts and then replace them with more realistic and supportive thoughts. The goal of CBT is to learn skills to change patterns of thinking or behaviors that are behind a person's difficulties, and consequently, help them change the way they feel. CBT is *problem-focused* and *action-oriented*. Clients learn specific skills that they can use for the rest of their lives. These specific skills help people identify distorted thinking, modify beliefs, relate to others in different ways, and change behaviors.

There are various cognitive–behavioral techniques that are widely used. Some of these techniques are as follows.

> Panic control therapy (PCT) is a widely used, empirically validated cognitive–behavioral treatment initially developed for the treatment of panic disorder, limited agoraphobic avoidance, and in combination with a situational exposure component. It is used as an aid to pharmacotherapy discontinuation in panic/anxiety disorders.
>
> Assertiveness training is another form of behavioral therapy designed to help people stand up for themselves. It is based on the idea that assertiveness is not inborn but a learned behavior. Behaving assertively can be difficult for almost anyone. By learning how to be assertive, the individual learns the appropriate balance between passivity and aggression. Assertiveness training is built on the principle that we all have the right to express our thoughts, feelings, and needs to others, as long as we do so in a respectful way. When we feel we are not able to express our feelings openly, it may lead us to depression, anxiety, and anger; and our sense of self-worth might suffer.

- Mindfulness-based approaches have their roots in ancient Buddhist traditions such as Vipassana and Zen meditations. Mindfulness is often incorporated into other therapeutic modalities. These are mindfulness-based cognitive therapy (MBCT), dialectical behavioral therapy (DBT), and acceptance and commitment therapy (ACT).
- Interoceptive exposure is also a cognitive–behavioral therapy originally used in the treatment of panic disorders; however, it is currently used for any problem where normal body sensations are experienced as threatening.

Neuropsychological Evaluation

This is an assessment conducted by a neuropsychologist or a trained skill psychometrist. Neuropsychological evaluations document patterns of strengths and weakness among cognitive and behavioral functions, behavioral alterations, and cognitive changes resulting from a central nervous system (CNS) disease or injury. Neuropsychological tests evaluate functioning in a number of areas including intelligence, executive functions (such as planning, abstraction, and conceptualization), attention, memory, language, perception, sensory motor functions motivation, mood state and emotion, quality of life, and personality styles. The test results of these tests help to understand more about how the brain is functioning.

RELAXATION THERAPY

This type of therapy encompasses many techniques. "A relaxation technique is any method, process, procedure, or activity that helps a person to relax; to attain a state of increased calmness; or otherwise to reduce levels of pain, anxiety, stress or anger" (Bourne, 2015). The health benefits of relaxation techniques include and are not limited to a decrease in muscle tension, cardiac health (i.e., lower blood pressure; slower heart and breathe rates), headaches, and insomnia. Meditation, for example, has been used for more than 3,000 years for the purpose of training and calming the mind. It was originated as a spiritual practice within the Hindu and Buddhist cultures. Some of the methods are performed alone while some require help from another person (often a trained professional). Some involve movement, some focus on stillness, and other methods involve the different elements.

Among the relaxation techniques that include normal and passive relaxation exercises are biofeedback, deep breathing, meditation, Qigong, self-hypnosis Zen yoga, yoga nidra, transcendental meditation, visualization, hypnotherapy, and pranayama.

Movement-based relaxation methods include some form of body work. This includes *exercise*, such as walking, gardening, yoga, and t'ai chi. Other examples include massage, acupuncture, reflexology, Feldenkrais, myotherapy, and self-regulation.

Some other relaxation techniques that can be used in conjunction with other activities include autosuggestion, prayer, and listening to new-age and classical music, all of which increase feelings of peacefulness and a sense of ease.

EYE MOVEMENT INTEGRATION (EMI) THERAPY AND EYE MOVEMENT DESENSITIZATION AND REPROCESSING (EMDR)

These are techniques that have the client recall the traumatic memory while visually following the practitioner's fingers through a series of eye positions that include eye-accessing cues. Having the client access the eye-access cues and memory simultaneously by adding eye-access positions from other sensory systems will allow the individual to change the sequence in which the person accesses the traumatic memory. It alters the sensory memory of the trauma. The client will re-experience the sensory elements from a traumatic experience in a different sequence, deleting the content from one or more sensory systems.

MEDICATION

Many accidents resulting in traumatic injuries are triaged by the 911 emergency responding team (ambulance or helicopter) and treated appropriately at hospital emergency departments. Trauma centers work closely with their respective EMS systems so care begins prehospital.

Trauma guideline alerts in the United States were established in 1976. These determinations will help professionals assess victims' immediate medical needs and transport them to the most appropriate hospital. Level I centers have optimum capacity through specialized staff and resources around the clock; these facilities provide the highest level of care available. United States are accredited and designed as Level I, II, III, or IV alert centers, based upon the care and volume they are able to provide.

Some of the most common traumatic injuries include, but are not limited to, spinal cord and brain injury, electrical injury, spine fractures, crush injury, broken bones, collapsed lungs, facial trauma, concussion, skull fracture, amputation, hypovolemic shock, subdural hematoma, and burns. The injury severity score (ISS) classifies each injury in every body region and according to the level of severity.

The type and length of medication given to the injured will vary according to the level of severity of injury. Once medical stabilization has been achieved, the prescription and use of *psychotropic* medications (to treat mental health–related disorders) also may be appropriate and used in conjunction with other types of medications following a trauma event.

It is recommended that a physician, preferably a knowledgeable and experienced psychiatrist, provide the treatment. For the purpose of this chapter we refer to anxiety disorders; depressive disorders; trauma and stressor-related disorders; and major or mild cognitive disorders due to a traumatic brain injury, substance-related disorders, or addiction-related disorders. The following are some of the most widely used medications:

- SSRIs antidepressant medications (selective serotonin reuptake inhibitor)
- High-potency benzodiazepines
- Tricyclics antidepressants
- MAO-inhibitor antidepressant
- Beta blockers

When the above medications are ineffective, psychiatrists may try other medications in conjunction with the above, including Depakote, Neurontin, Gabitril, or Lyrica, although anti-seizure medication have yet to be proven effective in treating anxiety disorders.

PHYSICAL THERAPY

Physical therapy is needed when a person experiences health problems that make it hard to move around and perform everyday tasks. Treatment is based on the findings of an initial evaluation and it may consist of a variety of hands-on joint and soft treatment techniques, exercise, and pain-reducing modalities. These include traction, ultrasound, electrical stimulation, and so on. Treatments are designed to treat injuries or illnesses, increase motion and strength, reduce pain, and, most importantly, restore function. They are also used as preventive method for future health problems.

ATTENDANCE OF SUPPORT GROUPS

Grief and bereavement groups have been increasingly recognized as an effective way to promote healing through education and support after the loss of a loved one. Research suggests that while there may not be scientific evidence to support its benefits, participants report strong positive outcomes, ranging from the psycho-educational learning component to the safe environment that encourages the open expression of feelings. The individual can tell his or her story and express feelings freely, knowing that others will be understanding, nonjudgmental, and supportive. Participants report that these groups serve as major stepping stone along the path of learning how to live with loss. Friendships are formed and these new support systems aid members' transition from being focused in the past to focus on the here and now and future.

Taking into consideration the complexities and uniqueness of the recovery process for each individual, the fact that ongoing evaluations and treatments are deemed necessary and that treatment can last a few months to years before an individual fully recovers, attorneys must consider the psychological component of trauma as an important factor when negligence is being alleged or when attempting to defend allegations of negligence.

REFERENCES

American Psychiatric Association. 2013. *Diagnostic and Statistical Manual of Psychiatric Disorders (DSM-V)*. Arlington, VA: American Psychiatric Association.
Bourne, E. 2015. *The Anxiety and Phobia Workbook*. California: New Harbinger Publications.
Goleman, D. 1996. Relaxation surprising benefits detected. *The New York Times*. Retrieved May 23, 2006.
Grollman, E.A. 1977. *Living When a Love One Has Died*. Boston; Beacon Press.
Hernan, J. 1992. *Trauma and Recovery*. New York: Basic Books.
Holland, D. 2011. *Grief and Grieving*. New York: Penguin Group.
James, J.W., and Friedman, R. 2009. *The Grief Recovery Handbook*. New York: Harper Collins.

Kübler-Ross, E., and Kessler, D. 2005. *On Grief and Grieving: Finding the Meaning of Grief through the Five Stages of Loss*. New York: Scribner.
Kushner, H. 1984. *When Bad Things Happen to Good People*. New York: Random House.
Robinson, L., Segal, R., Segal, J., and Smith, B. 2011. Relaxation techniques for stress relief. Retrieved from http://helpguide.org.
Smith, J.C. 2007. The psychology of relaxation, in eds. P.M. Leher, R.L. Woolfolk, W.E. Sime, *Principles and Practices of Stress Management*, 3rd ed.

WEBSITES

http://www.adaa.org: Anxiety and Depression Association of America
http://www.apa.org: American Association Help Center
http://www.becksinstitute.org: Anxiety and depression information
http://www.cancer.net: Coping with cancer/managing emotions/coping with guilt
http://www.grief.net: Helping people move beyond loss
http://www.helpguide.org: Information on how to cope with trauma
http://www.livingwithloss.com: Bereavement magazine
http://www.namy.org: National Alliance on Mental Illnesses
http://www.nimh.nhi.gov: National Institute of Mental Health
http://www.psych.org: American Psychiatric Organization
http://www.psychcentral.com: Information on psychological conditions
http://www.psychologytoday.com: Various articles on mental health
http://www.ptsdinfo.org: Information on PTSD
http://www.webcrawler.org: Tips to deal with PTSD

13 Drowning Forensics, Drowning and Accident Reenactments, and Case Research

John R. Fletemeyer

CONTENTS

Essentials of a Forensic Investigation 250
    Stages and Description of Body Decomposition 250
    Initial Assessment 252
    Risk–Benefit Factor 252
    Initial Underwater Survey 252
    Search for Victim(s) 253
    Evidence Recovery 253
    Victim Recovery 253
    Other Considerations 254
    The Diatom Test 254
    Other Drowning Tests 255
    Determining the Timeline for a Fatal Drowning 256
Drowning and Accident Reenactments 256
    Diving Reenactments 257
    SCUBA Reenactments 257
    Boat Injury/Drowning Reenactments 258
    Investigation and Research 258
Case Studies 258
    Low-Head Dam Drowning 258
    Shallow Water Diving Accident 258
    Large Dam Boat Accident 259
    Swimming Pool Drowning 259
References 260

Evaluating a fatal drowning event is often regarded as one of the most difficult challenges among forensic experts. Opinions from autopsy findings often contradict and may lead to multiple interpretations and erroneous conclusions, sometimes ones that are misleading. Gannon and Gilbertson (2014), after reviewing the results of several autopsies that were incorrectly reported as an accident, state:

> What was really disturbing was that I always assumed that a forensic pathologist was almost infallible when it came to medical diagnosis. Of course, after reviewing numerous autopsies and examinations over the years, I have found much to my dismay that this isn't correct.

Becker (2004) states that one myth about drowning is that they are presumed to be accidental. Based on his experience, this is often not the case. Many drownings are not accidental but in fact are involved in a homicide. Many fatal drownings following criminal proceedings may end up in a civil court and require a follow-up by a second and more comprehensive forensic investigation.

ESSENTIALS OF A FORENSIC INVESTIGATION

Although artificially derived, the process of how a body decomposes in the water is often divided into five stages: Fresh, bloated, decay, post-decay, and skeletal. Describing decomposition in stages is of special value in court proceedings and because jury members seldom have any background in biological sciences (Goff, 2009). Progressing from one stage to the next is often effected by water temperature, water chemistry, water depth (pressure), the characteristics of bottom sediments, currents, clothing worn, body weight, and the types and abundance of flesh-eating and decomposing marine/aquatic fauna.

STAGES AND DESCRIPTION OF BODY DECOMPOSITION

1. Fresh Stage

    The fresh stage begins at the moment of death and continues until bloating of the body is evident. In the majority of cases, a body becomes negatively buoyant as a result of air escaping from the lungs and being replaced by water. The deeper the water, the more negatively buoyant a body becomes. Typically a body will remain underwater until bacterial decomposition creates a buildup of gas inside the body to make it buoyant. The timeline for this varies significantly depending on a number of factors, including water temperature, water depth, and the body mass of the victim. In a majority of cases, a body stays submerged for one to two weeks. However, in very deep water, a body may never become positively buoyant because the pressure will reduce the volume of gas required for the body to float to the surface. There are few distinctive, gross decomposition stages associated with the body during this stage, although greenish discoloration of the abdomen and liver, skin crackling, and tache noire may be observed.

2. Bloated Stage

The principle component of decomposition, putrefaction, begins during the bloated stage. The anaerobic bacteria in the gut and other parts of the body begin to digest the tissues. Their metabolic processes result in the production of gases that cause inflation of the abdomen. As this stage progresses, the body may assume an inflated, balloon-like appearance.

3. Decay Stage

In this stage the body deflates and the skin begins to break loose from the body.

4. Skeletal Stage

At this stage the flesh removes from the body and only the skeleton remains. Once the body becomes a skeleton, several reliable methods can be used to determine age. One method that can be used is a radiographic analysis of the skeleton's hand and wrist. The Grulich and Pyle (1994) method takes a radiograph of the victim's hand and wrist and then uses an atlas of hands and wrists of known ages and then determines which the closest fit is. Other age determination methods include dental eruption patterns, epiphyseal closure of long bones, tooth wear, and cranial suture closure. Some of these methods can be combined to further increase reliability. In regards to these methods, it must be recognized that they are of only limited value in regard to reconstructing the drowning event.

Complicating the reliability of forensic findings is how a body is recovered. Unless recovered by using properly established search-and-recovery methods, critical evidence may be lost. What makes evidence recovery problematic relates to the fact that aquatic environments are constantly changing. Consequently, recovery operations and investigations are seldom afforded the luxury of unlimited time to conduct work so out of necessity and safety, bodies must be recovered quickly.

There are established, professional guidelines when investigating a fatal drowning, especially if the drowning was not witnessed and suspicious circumstances are indicated. These guidelines are similar to the guidelines established for land-based investigations but with some important differences due to the challenge of recovering bodies and evidence in an aquatic environment.

There are several factors that complicate an aquatic crime scene investigation, including the following:

- Limited water visibility.
- Bottom sediments that make evidence recovery difficult and even impossible. A silt or mud bottom may quick engulf evidence.
- Bottom obstructions and entanglements such as boulders and human debris.
- Strong currents.
- Water contamination, that is, oil, gas, and sewer water.
- Marine life, that is, alligators, snakes, and snapping turtles.

Failing to follow professionally established guidelines may lead to a botched investigation and an unreliable interpretation of the victim's death. In *Case Studies in Drowning Forensics* (Gannon and Gilbertson, 2013) eleven cases were examined and determined that serious mistakes were made that led to conclusions of an accidental drowning. Instead upon further examination, in all eleven cases, forensic evidence had been overlooked or disregarded indicating a homicide. Gannon and Gilbertson reported that medical examiners often work without the benefit of medical training and that law enforcement officers are conditioned to regard drowning as "accidents," seldom considering the possibility of foul play.

The record is replete with cases where crucial evidence was not recovered and properly preserved. In several cases, court rulings prevented evidence from being introduced for jury consideration. This was based on acknowledgment that evidence had been contaminated and improperly documented by personnel lacking training in crime scene investigations and underwater search-and-recovery methods.

As a rule of thumb, nothing in the underwater scene should be touched, disturbed, or moved until documented with photographs, video, and sketches with accompanying notes and investigation reports. This should be conducted as a coordinated effort under the supervision of a designated crime scene supervisor and the search-and-recovery team leader.

The following represents the steps that should be taken when conducting a forensic recovery of a submerged victim.

INITIAL ASSESSMENT

During the initial assessment an accident scene supervisor and dive team leader will be designated. The initial assessment should be conducted during daylight hours and the immediate area should be secured as a crime scene including the likely entry points. Witnesses should be identified and intelligence collected. The area should be evaluated for safety hazards and conditions that might impact the investigation that will follow.

During the initial assessment, a safety evaluation should be conducted to prevent risks to staff, especially to dive team members.

RISK–BENEFIT FACTOR

The benefit of the evidence-and-recovery operation is weighed against the risk of the dive to determine if it is safe to dive. The six factors listed previously should be considered when weighing the risks.

INITIAL UNDERWATER SURVEY

An initial underwater survey is conducted to identify problems that impact the recovery operation. Determining the water visibility on and near the bottom, deifying of thermoclines and current velocity are important. Upon surfacing and returning to shore, a detailed report is made to the dive team leader in order to develop a plan for the operation.

Search for Victim(s)

Various search patterns must be considered based on some of the physical factors encountered and the area to be searched. Large search areas versus small areas dictate different search patterns and methods. If water depth exceeds 34 feet, decompression must be considered to ensure decompression limits are not exceeded.

When searching for a victim, divers must be as careful as possible not to disturb the area. Kicking up mud or silt can potentially destroy valuable evidence or prevent it from being seen. Once the victim has been found, the location of the body should be marked with a maker buoy and then the next phase of the search-and-recovery operation, evidence recovery, should begin.

Evidence Recovery

Each diver participating in the evidence recovery-and-documentation phase of the operation will be given assignments by the dive team leader. Noting and sketching the location and position of the body is of critical importance. If water conditions permit, underwater photographs should be taken. In addition, it is important to take a water sample in the event that a diatom test is warranted (refer below). In addition, the protocol for evidence collection in an underwater crime scene includes the following:

- Photograph and video tape the underwater scene, including from on land and underwater, if conditions permit.
- Describe the position of the body, water temperature, and thermoclines.
- Describe the bottom and take a sample of bottom sediment and several water samples using a water sample bottle—not just any bottle!
- Observe the condition of the body and signs of trauma.
- Observe clothing and anything unusual, such as rips in clothing, wearing shoes, jewelry, wallet, and so on.
- Bag the victims hands, feet, and head with socks, plastic bags, and so on.
- Wrap the victim from head to toe in a sterile white sheet and secure with duct tape, bungee cords, ties, and so on.
- Carefully bag the body using an approved body bag.
- Conduct a follow-up search to identify evidence. If there is a strong current, the search should be conducted well downstream.

Victim Recovery

Following the follow-up dive and evidence recovery, the victim is then brought to the surface and returned to shore. Upon reaching shore, the medical examiner takes control of the body with the assistance of paramedics to assist with transport.

Evidence is then turned over to crime scene investigators and evidence recovery specialists where it is cataloged and taken to a secure location such as an evidence locker or room. For more information about evidence preservation refer to Chapter 4 by St John.

Nothing in the crime scene should be touched or disturbed until documented with photographs with sketches (with reference points). Each incident is processed as a crime scene during the entire operation, or until the requesting law enforcement agency determines otherwise and requests a recovery operation only with no evidence collection.

Other Considerations

Failure to follow any of the measures described in the above provides the opportunity for a successful Daubert challenge, resulting in crucial evidence from being admitted into the courtroom. In addition, when a body is discovered in the water, there are several issues that must be considered and resolved.

- Is the postmortem period the same or different than what is stated by a witness or witnesses?
- Is there pathological evidence? This appears in the medical examiner's report and may include evidence of postmortem trauma inflicted upon the victim's body.
- Is there any evidence of natural causes that may be responsible, providing an answer to a drowning, or at least a contributing factor?
- Were drugs or alcohol involved? This will be revealed in a toxicology report.
- What was the psychological state of the victim? Was he or she depressed, suffering financial problems, or had recently experienced the death of a loved one? Regarding the victim's psychological state, sometimes it is necessary to add a psychologist to the forensic team to determine this.

The Diatom Test

Diatoms are unicellular algae belonging to the class of Bacillariophyceae, which includes more than 15,000 organisms living in fresh water, brackish water, and sea water (see Figure 13.1). The skeletons of these algae are called frustules and are made of hard silica. Several investigators independently discovered that diatoms can be recovered from decomposing bodies and that their presence in human tissue indicates whether the victim suffered a fatal drowning event or died on land. While actively drowning in an openwater environment (i.e., a lake, pond, river, canal, or ocean), a victim ingests water containing diatoms through the alveoli and upon microscopic examination of collected tissue samples, diatoms in varying numbers are usually identified.

Usually the objective of a drowning forensic investigation is to determine if the victim suffered a drowning event (death by submersion) or, alternatively, died prior to submersion. Historically, a diatom test where diatoms were identified in postmortem human tissue was considered evidence for drowning. However, recent research discovered that diatoms are often found in the tissue of non-drowned victims

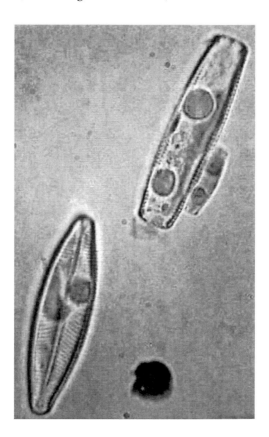

FIGURE 13.1 Diatoms are unicellular algae found in fresh water, brackish water, and sea water, but usually not in bathtub water or swimming pool water. The diatom test can be used to determine if the victim drowned (suffered death by submersion).

(Foged, 1983; Schellmann and Sperl, 1979; Schneider, 1980), thus creating a false positive for this test and making its validity suspect. However, additional and more recent research discovered that relatively large numbers of diatoms could, with confidence, indicate a fatal drowning. Auer and Mottonen (1988) reported that if a minimum of twenty diatoms are found in a slide, this represents a threshold quantity necessary to indicate that the victim drowned. Farrugia and Ludes (2006) reported a similar finding. This new research is responsible for re-establishing the validity of the diatom test but under certain conditions and quantitative thresholds. However, in certain aquatic environments (i.e., bathtubs and swimming pools), it does not serve any value as a diagnostic test for drowning because in these environments, it is unlikely that diatoms are present.

OTHER DROWNING TESTS

Several other forensic methods are used to evaluate a drowning. Large amounts of froth (often pinkish in color) present around victims' mouths and nostrils is usually an indicator of a drowning event. Blood froth is also commonly observed in the victim's lower and upper airways (Farrugia et al., 2006). However, if the drowning occurs where there is a current, this evidence can be washed away.

In some cases, higher-than-normal lung weights after becoming waterlogged determined during an autopsy represents an indicator of a drowning event, but not always; sometimes, hemorrhaging of the alveoli walls may indicate that the victim drowned.

DETERMINING THE TIMELINE FOR A FATAL DROWNING

Also challenging is determining a timeline for a drowning event. On land, this is fairly easy; in the water, it is challenging because so many variables are involved. On land, taking the core temperature of the deceased can be used to determine the time of death, because for every 10 minutes, body core temperature usually drops by about 1°F. In many cases, the body temperature will reach ambient temperature in 18 to 20 hours (Fisher 2007). Because water dissipates heat five times faster than air, this must be considered. Consequently, this makes this method a less reliable indicator. Nevertheless, it is always useful to take a postmortem core temperature and is recommended when a body is found in a "fresh" condition.

Skin color is also an indicator but changes at different rates depending on water temperature. For the initial six hours, the skin color remains in a fairly normal state until pooling of blood in the extremities results in an ashen color, especially in the face. Postmortem skin changes in skin color are less pronounced in a dark-skinned victim.

Another indicator used to establish a timeline is skin condition. Farrugia and Ludes (2006) report the following:

> One of the signs of immersion is skin maceration becoming visible after varying time intervals depending on the temperature of immersion water. The skin becomes wrinkled, pale, and suddenly like a "washer woman's" skin. These changes appear at the fingertips, palms, backs of the hands, and later, the soles. The next step is the detachment of the thin keratin of hands and feet, which pull off in "glove and stocking" fashion. Nails and hair become loosened after a few days. Other signs of immersion are cutis anserine and postmortem distribution of hypostasis. The presence of mud, silt, or sand on the body was described but has no diagnostic value.

Skin maceration begins quickly within the first 24 hours. As maceration progresses the skin becomes blanched, swollen, and wrinkled. It is first apparent in the skin of the fingernails and palms. In water 50 to 60°F early changes may be seen within an hour.

DROWNING AND ACCIDENT REENACTMENTS

In some potentially high-value civil cases, reenactments become an invaluable and effective tool, although often very expensive and always time-consuming to produce. In all instances, reenactments must be carefully planned and orchestrated. A poorly planned drowning or aquatic injury reenactment without a carefully stated objective may result in harming the case by contradicting or confusing the theory of negligence. In addition, it is important to plan reenactments that will withstand a possible Daubert challenge.

Diving Reenactments

Of all the types of aquatic injury and drowning reenactments, diving reenactment is perhaps the most valuable because it provides a jury visual information necessary to understand allegations of negligence. When performing a diving reenactment, the following factors must be considered:

- Victim's height and weight.
- The distance between the water and the surface that the dive was executed.
- The depth of the water.
- Water visibility.
- The contour and hardness of the bottom.
- If in the ocean, the tidal state at the time of the accident.
- If witnesses were present, the trajectory (angle) that the dive was executed.

Whenever possible, a diving reenactment should include both land-based photo documentation as well as underwater documentation.

SCUBA Reenactments

SCUBA reenactments are challenging because SCUBA drownings and accidents often are not witnessed. In addition, they involve compressed air and the associated dangers with compressed air injuries, that is, air embolism (see Figure 13.2). Reenactments should never be conducted without a safety diver.

FIGURE 13.2 SCUBA diving reenactments although challenging provide useful jury exhibits.

Boat Injury/Drowning Reenactments

In most cases, the same method commonly used to reenact vehicle accidents can be used to reenact a boating accident. There are literally hundreds of services throughout the United States that provide this service.

Investigation and Research

Research is often conducted to prove or disapprove a theory of negligence and must always consider the possibility of being challenged by a Daubert test (refer to Chapter 4 by St John). Therefore, research must not only be carefully designed to be reliable and valid but also amenable to replication by opposing council experts. At the time of this book's publication, several research investigations have been conducted in response to serious aquatic accidents or fatal drownings. The following provides a few examples of research conducted by the author describing the research method, objectives, and findings.

CASE STUDIES

Low-Head Dam Drowning

Objective: To demonstrate the formation of hydraulics and the difficultly of escaping hydraulic currents commonly found immediately below a low-head dam.

Methods: Current dyes were injected into dam currents upstream and current velocities were taken at various locations. A wave tank study was conducted to demonstrate how currents are created by a low-head dam.

Results: This research identified hazardous currents resulting from the location of the dam, making it difficult if not impossible for an average swimmer to escape. In addition, this research found that hazardous currents created by the dam could not be detected by making casual observations and without the benefit of warning signs.

Shallow Water Diving Accident

Objective: To demonstrate that the shallow depth of the water was responsible for the spinal injury sustained by the victim.

Methods: The depth at the time of the accident was determined after performing a topographic profile study of the bottom. A cone penetrometer was used to determine the hardness of the bottom. A Secchi disk was used to determine the water visibility at the accident scene. Several reenactment dives were performed under control conditions to determine the velocity and trajectory of the victim striking the bottom and sustaining a spinal injury.

Result: The shallow water and performing a dive from a boat dock at night was the proximate cause of the victim's spinal injury. The victim was aware that this was a boat dock and not a platform where a dive could be safety executed, especially at night when it was not possible to see the sea bottom.

FIGURE 13.3 A drone was used to document a dummy victim being swept over a dam.

LARGE DAM BOAT ACCIDENT

Objective: To determine the velocity of the current flowing over a diversion dam and that velocity was sufficient to capture a boat and cause it to be swept over the dam.

Methods: Drift buoys and a "dummy" victim were placed in the dam's overflow current (see Figure 13.3). Current dyes were injected into dam's current and a drone was used to observe the current overhead.

Result: The power and velocity of the current created by the overflow water was sufficient to sweep a small boat and the victim over the dam. Even if rescuers were present, it is unlikely that the victim could have been rescued. The presence of the dam and other visual cues were sufficient to warn the victim about the dangers of being close to the dam. The operator's operation of the boat within close proximity of the dam was the proximate cause of the accident and fatality.

SWIMMING POOL DROWNING

Objective: To determine if a failure to monitor pool chemistry was responsible for the poor water visibility that contributed to a drowning in the pool's deep end.

Methods: A Secchi disk was used to determine water visibility and a study was conducted to demonstrate how improperly treated pool water created poor water visibility.

Results: Results of the experiments indicated that improperly treated water was responsible for the pool's poor water visibility and was a contributing factor to the drowning event. The lifeguards closed down the pool's deep end and allowed swimming in the shallow end. This action was a violation of swimming pool standards and was a contributing cause of the fatal drowning event.

REFERENCES

Auer, A., and Mottonen, M. (1988). Diatoms in drowning. *Z. Rechtsmed* 101:87–98.

Beckor, R. (2000). Myths of underwater recover operations. *FBI Law Enforcement Bulletin* 69(9):1–5.

Farrugia, A., and Ludes, B. (2006). *Diagnostic of Drowning in Forensic Medicine*. Institute of Legal Medicine, pp. 1–24.

Fisher, B. (2007). *Techniques of Crime Scene Investigation*, 7th ed. New York: CRC Press.

Foged, N. (1983). Diatoms and drowning—Once more. *Forensic Science International* 21:153–159.

Gannon, K., and Gilbertson, D. L. (2014). *Case Studies in Drowning Forensics*. New York: CRC Press.

Goff, M. L. (2009). Early post-mortem changes and stages of decomposition of exposed cadavers. *Experimental Science and Application* 49:21–26.

Greulich, W., and Pyle, S. I. (1994). *Radiographic Atlas of the Skeletal Development of the Hand and Wrist*, 2nd ed. California: Stanford University Press.

Ludes, B., Coste, M., and Mangin, T. (1999). Diatom analysis in victim's tissue as an indicator of the site of drowning. *Journal of Legal Medicine* 112:163–166.

Piette, M. H. A., and De Letter, E. A. (1966). Drowning: Still a difficult autopsy diagnosis. *Forensic Science International* 163:1–9.

Polamen, M. S. (1998). *Forensic Diatomolgy and Drowning*. Amsterdam: Elsevier.

Schellmann, B., and Sperl, W. (1979). Nachweis in Knochenmark Nichtertrunkener. *Z. Rechstmed* 83:319–324.

14 Boating and SCUBA Accidents, Low-Head Dams, and Beach Renourishment

John R. Fletemeyer

CONTENTS

Introduction	262
Boating Accidents	262
Boat Safety Equipment	263
Boating under the Influence (BUI) of Alcohol	263
Inexperience	265
SCUBA Accidents	265
Dive Training	266
Accidents	266
Diving Hazards	267
Sea Life	268
Decompression Sickness	268
Oxygen Toxicity	269
Barotrauma	269
Nitrogen Narcosis	269
Pulmonary Embolism	269
Defective Equipment	269
Shallow Water Blackout (SWB)	270
Snorkeling	270
SCUBA Dive Charters and Snorkeling Trips	271
Shark Diving Encounters	272
Low-Head Dams	274
Beach Renourishment	278
References	279

INTRODUCTION

The objective of this chapter is to provide members of the legal profession information about three categories of aquatic accidents that are often involved in a suit alleging gross negligence. Each of these categories often involves several contributing and mitigating factors when evaluating the level of negligence involved. Consequently, determining the root or proximate cause(s) of an accident is often challenging, even for an experienced attorney. Regarding these three unrelated categories of accidents, it is only possible to provide an overview of each. Additional resources may need to be consulted before developing a theory of negligence or identifying an appropriate defense response given the set of circumstances surrounding the accident.

BOATING ACCIDENTS

According to U.S. Coast Guard statistics, there are more than 12 million registered boaters. On average, between 7,000 and 8,000 boating accidents are reported every year. It is estimated that 70 percent of boating accidents are caused by operator error as opposed to a boat or environmental factor.

Legally, there are three scenarios that are defined as boating accidents. It is important to note that accidents may or may not involve a collision. The first scenario is when a vessel is damaged more than a minimum amount, which varies in every state. The second scenario is when a boat passenger dies or is seriously injured. The third scenario is when a boat passenger disappears from the boat and it is suspected that death or injury has occurred. In all three of these cases, the operator or owner of the boat is legally required to file an accident report with appropriate state or local authorities or the Coast Guard.

As Americans have more leisure time to devote to water activities and as they have more disposable income resulting from an improving economy, boating will gain popularity. In addition, there is a growing trend of bigger and more powerful boats (Sweers, personal communication). Consequently, bigger and faster boats create greater opportunities for serious accidents, especially involving new and inexperienced boat owners.

Operating a boat, especially on a crowded waterway or lake, represents an enormous responsibility that falls mainly on the boat owner or the person operating the boat, who henceforth will be referred to as *captain*. The captain is responsible for the safe and efficient operation of the boat. All souls on board, including crew and guests, are under the captain's authority. Maritime law is clear about this with an abundance of case law holding the captain responsible for passenger safety. This is mandated not only by the International Maritime Organization, but also by the United States Coast Guard as well as many maritime authorities, including the Marine Patrol, DNR, U.S. Fish and Wildlife, and local enforcement agencies. There are only a few circumstances where the courts have ruled that a boat captain was not held accountable for the safety of his or her boat and its passengers. For example, courts have ruled that a captain is not responsible when a weather-related event was proven responsible for an accident.

Regarding the captain's responsibility, the U.S. Coast Guard states that

> A ship's captain commands and manages all ship's personnel, and is typically in charge of the boat's accounting, payrolls, and inventories. The captain is responsible for compliance with immigration and customs regulations, maintaining the ship's certificates and documentation, compliance with the boat's security plan as mandated by the International Maritime Organization. Most importantly, the captain is responsible for responding to and reporting in case of accidents and incidents and is responsible for the safety of the boat's crew and passengers.

It should be noted that a captain's responsibility for passenger safety not only occurs while under sail but also at times when the boat is moored. In addition, when boats are involved with a business, the captain's responsibility for passenger safety is increased.

BOAT SAFETY EQUIPMENT

The U.S. Coast Guard requires pleasure power boats less than 65 feet to be equipped with certain safety items, including:

- Personal flotation devices (PFDs). One type of Coast Guard-approved life jacket must be on board for each person on the boat. Also, each boat must have one Type V, or throwable-type PFD.
- Visual distress signal. One orange distress flag and one electric distress light; or three handheld or floating smoke signals and one electric distress light; or three combination red flares, handheld, meter, or parachute-type.
- If the boat was built after 1980 and uses gasoline, it must have an exhaust blower.
- If the boat is operated after sunset and before sunrise, it must have navigation lights.
- Fire extinguisher. One marine-type USCG B-1 or three B-1 fire extinguishers if boat has an inboard engine and enclosed compartments where fuel or flammable and combustible materials are stored.

In addition, smaller craft such as wave runners should have an engine kill switch that can be attached to the operator's wrist. In the event that the operator falls from the craft and into the water, the craft's engine will immediately turn off, preventing a runaway craft and a serious accident.

BOATING UNDER THE INFLUENCE (BUI) OF ALCOHOL

Alcohol is the most important cause of serious boating accidents and is arguably the most preventable. In more than one-third of boating accidents, a driver or captain who is under the influence of alcohol is a key contributing factor. Boating under the influence (BUI) is a criminal offense is every state.

As little as one alcoholic drink is enough to impair the judgment of a captain. The U.S. Coast Guard states the following about the dangers of BUI:

- A boat operator is more likely to be impaired than a driver.
- The penalties for BUI can include large fines, revocation of operator privileges, and serious jail terms.
- The use of alcohol is involved in about one-third of recreational boating fatalities.

Alcohol affects judgment, vision, balance, and coordination. These impairments increase the likelihood of accidents afloat—for both passengers and boat operators. U.S. Coast Guard data shows that in boating deaths involving alcohol use, over half of the victims capsized their boats or fell overboard. Alcohol is even more hazardous on the water than on land. The marine environment—motion, vibration, engine noise, sun, wind, and spray—are known factors that accelerate a drinker's impairment. These stressors cause fatigue that makes a boat operator's coordination, judgment, and reaction time decline even faster when using alcohol. Alcohol can also be more dangerous to boaters because boat operators are often less experienced and less confident on the water than on the highway. Recreational boaters do not have the benefit of experiencing daily boat operation. In fact, data reports that boaters on average have only 110 hours of time on the water per year.

Demonstrated effects of alcohol include the following:

- Cognitive ability and judgment deteriorates, making it harder to process information, assess situations, and make good judgment.
- Physical performance is impaired—evidenced by balance problems, lack of coordination, and increased reaction time.
- Vision is affected, including decreased peripheral vision, reduced depth perception, decreased night vision, poor focus, and difficulty distinguishing colors.
- Alcohol creates a physical sensation of warmth, which may prevent a person in cold water from getting out before hypothermia sets in.

Many factors, including prescription medications and fatigue, can affect an individual's response to alcohol, and impairment can occur much more rapidly as a result. There is no safe threshold for drinking and operating a boat.

Many boating accidents are caused by adverse weather, such as strong winds or heavy rain. Weather is especially dangerous when it comes up suddenly and unexpectedly. In these cases, even experienced boaters may have trouble navigating, avoiding collisions, and, in rare cases, keeping the boat afloat. In bad weather, a boat may be exposed to a lightning strike, which can injure or kill a passenger. In addition, exposure to cold water or rain can lead to hypothermia, especially in boats with open cockpits.

INEXPERIENCE

Fewer than 30 percent of boat owners have taken a safe boating class. In addition in most states, minors are allowed to operate a power boat. However, when minors operate a power boat, the horsepower usually may not exceed a certain limit—usually less than 12 horsepower in some state and less than 25 horsepower in other states.

Inexperienced boaters are at a higher risk of being involved in a boating accident. An inexperienced boater may have more trouble controlling the boat in rough water and may demonstrate bad judgment in dealing with situations potentially impacting the safety of the passengers. An inexperienced boater may be confused when navigating crowded waterways and about avoiding dangerous areas. Many more inexperienced boaters run aground than experienced boaters. For these reasons, it is critical that inexperienced boaters practice boating and navigation skills. In addition, a new boat owner who is inexperienced should take a U.S. Coast Guard–sponsored safe boating course. According to Coast Guard data, fewer than 30 percent of boat owners have taken a boating safety course.

SCUBA ACCIDENTS

Diving in this chapter is defined as any dive involving the use of compressed air. Of the several types of diving, the use of SCUBA is the most common. SCUBA is an acronym for *self-contained underwater breathing apparatus*. SCUBA diving is considered an extreme sport, requiring specialized equipment and training. SCUBA diving was co-invented and made popular by Captain Jacques Cousteau in the late 1950s and early 1960s. SCUBA is a relatively safe activity. The Divers Alert Network (DAN) estimates that 1 out of every 211,864 dives ends in a fatality. Most forensic investigations of diving accidents, including fatal accidents, indicate that most are preventable. SCUBA diving fatalities occur in several different ways. Of the many ways that SCUBA divers die, drowning, air embolism, and cardiac arrest represent the leading causes.

Determining the number of SCUBA divers in the United States is difficult for several reasons. The major diving organizations, including PADI, NAUI, and SSI, are reluctant to publish membership data and there is still not a consensus opinion about how a diver should be defined. According to an article appearing in *Under Currents*, estimates range from 1 million to 3 million. The most accepted estimate is 1.2 million with a +/– range of 15 percent. The majority of divers are classified as sport divers, defined as a recreational diver who dives three or more times a year.

SCUBA diving requires that the participant be in good medical health. A clearance in writing from a physician is usually required before an individual is allowed to participate in any of the training programs that are conducted in every state. Other written clearances are also required and may represent a contributing factor in a negligence suit if not properly written. When defending a negligence suit, a signed clearance is often the first response.

In addition, an important prerequisite is that the participant is a competent swimmer. In most cases, the beginning of a SCUBA certification program involves a swimming test to determine whether a student has the necessary skills to partake in SCUBA and its rigors. In the event that a student is not tested or does not have the necessary skills, it is probable that a SCUBA instructor would be found grossly negligent if a student sustained a dive-related accident.

As more and more SCUBA divers are traveling to exotic places to dive, the industry is facing a problem in determining the competency of a diver. In many cases checking to see if a diver has an active certification card is not enough. Consequently, many SCUBA operations require divers to physically demonstrate their basic skills in a pool or in shallow water before allowing them to partake in a dive. Often a competency demonstration includes clearing a mask, clearing a regulator, and buddy breathing.

Another problem when diving in an exotic location involves the quality of air when tanks are filled at a local compressor. Most countries are not regulated to have air intakes and compressor filters changed regularly. Consequently, compressed air may be contaminated and may cause a serious problem, especially when diving at deep depths. As the pressure increases, the toxic effect of contaminated air also increases.

Dive Training

There are several dive training organizations in the United States. The two most respected include the National Association of Underwater Instructors (NAUI) and the Professional Association of Underwater Instructors (PADI). Other training organizations include the National Academy of SCUBA Educators (NASE), the SCUBA Schools International (SSI), and the Professional SCUBA Association International (PSAI). All of these training organizations have rigid training standards that usually follow recommendations from the World Recreational SCUBA Training Council (WRSTC) and from the International Organization for Standardization (ISO). The Diver Alert Network (DAN) is a highly professional not-for-profit organization that identifies standards pertaining to diver safety.

Accidents

According to the DAN, the top three causes responsible for diving fatalities are

1. Pre-existing disease or pathology in the diver.
2. Poor buoyancy control.
3. Rapid ascent/violent water movement.

There are various levels of training, including recreational diving. Some organizations now offer a resort diving course as the entry level. At this level, the recommended depth limit is 33 feet in salt water and 34 feet in fresh water, equaling one atmosphere of pressure.

Regarding accidents, SCUBA accidents are most common among novices, usually individuals with very little training and only a few open-water dives. Many accidents occur among individuals who have placed a temporary halt on diving and then suddenly resume, responding to a trip to an exotic location. In order to be a safe and confident diver, it is incumbent to practice skills and partake in refresher dive(s) involving an instructor. Many organizations offer refresher courses, and a refresher course is highly recommended if a diver has not been diving for a while.

DIVING HAZARDS

Several hazards exist that may be responsible for a serious accident or a fatality. Some of the most common and preventable accidents include the following:

- An improper entry resulting in a spinal or head injury.
- Failure to practice the "buddy" system (see Figure 14.1).
- Failure to exercise care when surfacing, resulting in striking the hull or prop, which leads to a serious injury or drowning. There have been several cases when a diver (SCUBA and snorkeling) surfaced beneath a boat and was rendered unconscious. Consequently, a fatal drowning occurred.
- Improperly storing or handling tanks on a boat may result in an injury in the event that a tank accidentally tips and falls.
- Failure to have a quick-release buckle in a position allowing for an immediate release in the event of an emergency assent.
- Exceeding no decompression limits defined by Navy dive tables.
- Breath-holding compressed air when surfacing.
- Failing to remain within visual eyesight of dive buddy.
- Failure to have an emergency dive plan and failure to practice the plan.
- Diving in a strong ocean current.
- Diving in low visibility water.

FIGURE 14.1 The buddy system is an important practice that promotes dive safety not only among SCUBA divers but also among snorkelers.

SEA LIFE

Dangerous marine life represents a threat to SCUBA divers. It is important to learn how to recognize dangerous marine life and how to avoid being stung or bitten. The danger of most marine life has been exaggerated, for example, the risk of being bitten by a shark. However, when dangerous marine life is encountered, it should always be treated with care and respect.

DECOMPRESSION SICKNESS

Decompression sickness (sometimes called "the bends") results from inadequate decompression and nitrogen bubbles coming out of suspension in the bloodstream. The DAN estimates that approximately 1,000 U.S. divers suffer from decompression sickness every year. Outcomes can range from minor (temporary headache and confusion) to major (paralysis and even death).

Decompression risk factors include the following:

- Deep and long dives
- Cold water
- Hard exercise at depth
- Rapid ascent
- Age

Decompression sickness symptoms include the following:

- Unusual and severe fatigue
- Itchy or blotchy skin
- Pain in the joints, arms, and legs
- Dizziness, vertigo, and ringing in the ears
- Numbness, tingling sensation, and paralysis (total or partial)
- Shortness of breath

Decompression sickness signs include the following:

- Skin appearing as being blotchy and sometimes red
- Paralysis and muscle weakness
- Behavior change (combative) or bizarre behavior
- Amnesia
- Difficulty urinating
- Collapse and unconsciousness
- Coughing up bloody, frothy sputum

A diver increases exposure to decompression sickness when performing several dives within a twenty-four-hour period or when diving immediately following a long flight in a pressurized air plane. Age, cold water, and level of exertion represent other contributing factors to decompression sickness.

OXYGEN TOXICITY

Oxygen toxicity is usually a problem only to deep divers who go below 135 feet. Like nitrogen, the body absorbs extra oxygen under increased underwater pressure as well. For most divers this is not a problem, but at extreme depths so much extra oxygen is absorbed that the life-giving gas becomes toxic. The effects range from tunnel vision or nausea to twitching to loss of consciousness or seizures.

BAROTRAUMA

Barotrauma (sometimes called *squeezes*) is caused by damage done by increased underwater pressure on the air pocket in the middle of the ear. Divers usually "equalize" during a dive by pinching their nose and blowing, by chewing, or by swallowing to push more air into the middle ear. However, a descent that is too rapid can overcome a diver's ability to equalize and result in severe pain and even injury to the middle ear. Because the greatest pressure occurs at shallower depths, squeezes are most likely to occur near the surface.

NITROGEN NARCOSIS

Another nitrogen-related danger is the narcotic effect that extra nitrogen has on the body. Anyone who has had nitrous-oxide gas at the dentist is already familiar with this effect. Nitrogen narcosis is a danger because it impairs judgment and sensory perception. As with the bends, the degree of nitrogen narcosis is related to how deep a diver goes and how much nitrogen is absorbed into the body.

PULMONARY EMBOLISM

Air embolism is one of the most dangerous and life-threatening of all the hazards encountered by a sport diver. Air embolism typically results when a diver breathing compressed air forgets and holds their breath while ascending to the surface. This results in the lungs expanding and sometimes causing an emboli or air bubble to enter into a vein or artery. In the worst cases, the bubble is lodged in the brain or near the heart muscle. When this happens, and without the benefit of immediate recompression, death often results.

DEFECTIVE EQUIPMENT

Many casual divers do not own their own equipment, and are therefore reliant on renting equipment from the SCUBA diving operator who is conducting their dive trip. A broken depth gauge could lead to a mild case of decompression sickness, while a bad regulator might result in drowning. A diver should always thoroughly check rented SCUBA diving equipment and never be shy about asking for a new piece of gear if they suspect something is wrong.

SHALLOW WATER BLACKOUT (SWB)

SWB is attributed to more and more diving fatalities, especially among free divers and snorkelers. SWB is the loss of consciousness caused by cerebral hypoxia toward the end of a breath-holding dive in water that is typically shallower than 16 feet when the swimmer does not necessarily experience an urgent need to breathe and has no obvious medical condition that might have caused it.

Because SWB is a complicated physiological event, most novice divers do not fully understand or appreciate the danger of hyperventilating several times before submerging. SWB shows no overt behavioral signs that a victim is drowning because before surfacing, the victim becomes unconscious.

Shallow water blackout can be easily prevented by not performing consecutive hyperventilations and by not engaging in breath-holding contests.

SNORKELING

Snorkeling involves the use of mask, fins, and a tube extending from the mouth to the surface called a "snorkel." In most cases, snorkeling is much safer than SCUBA diving. However, there is an important exception when a snorkeler is engaged in prolonged breath-holding activities. The following represents a list of factors that contribute to SWB:

- An individual who is out of shape and unable to withstand the rigors of free diving.
- An individual who has a heart condition.
- An individual who has been drinking.
- An individual who is not a good swimmer and who does not feel at ease in the water (see Figure 14.2).
- An individual who has not been properly trained or versed in using snorkeling equipment. It is necessary for a snorkeler to learn how to clear a snorkel. This cannot be learned by classroom instruction without practice, which should take place preferably in a swimming pool.
- Wearing improperly fitting equipment, especially a leaky mask.
- An individual hyperventilating several times, placing them in danger of SWB.
- Diving in rough water.
- Diving in a strong current.
- Diving from a boat and inadvertently surfacing under the boat, striking either the hull or propeller.
- Failing to understand and not knowing how to remedy barotrauma injuries, such as the squeezes, as discussed previously.

Boating and SCUBA Accidents, Low-Head Dams, and Beach Renourishment   271

FIGURE 14.2   Snorkeling injuries are most common among inexperienced bathers.

SCUBA Dive Charters and Snorkeling Trips

Several legal actions have been taken against boat excursions and snorkeling trips primarily designed for the novice or sport diver. The burden of diver safety primarily falls on the captain while divers are on the boat and on the dive master once divers are in the water (see Figure 14.3). Prior to entering the water, it is the dive master's and captain's responsibility to

- Check water conditions, including sea state, currents, and visibility to make sure that it is suitable for diving.
- Make sure that divers hold current certifications, and in the case of snorkelers, make sure that they have appropriate swimming skills suitable for existing diving conditions.
- Make sure that divers previously have not been engaged in drinking.
- Make sure that each diver has a buddy and will practice the buddy system.
- Review above- and underwater hand communications.

FIGURE 14.3 Dive trips are popular, especially among novices, and are hazardous when the crew is improperly trained, including the captain and dive master.

- Review emergency plan.
- Review the predetermined dive plan.
- Review boat entry and exit methods.
- Monitor divers for signs of stress and anxiety when entering the water; if signs are detected, deny entry.

At the completion of the dive, the captain or his or her designate must perform a headcount, making sure that all souls are accounted for. There have been several cases where a diver has been left at sea following the conclusion of a dive or snorkeling trip.

SHARK DIVING ENCOUNTERS

Shark encounters are increasing in popularity, and in the United States, they primarily occur in Florida and California. Shark encounter programs are also popular in several other countries, including in the Bahamas, throughout the Caribbean, Mexico, South Africa, Costa Rica, Australia, and countries bordering the Red Sea. These are tropical countries where sharks tend to thrive and where large populations inhabit coastal waters. See Figure 14.4.

Some programs feature shark cages, creating a protective barrier between the diver and the shark; other programs encourage free-swimming encounters using either SCUBA or snorkeling equipment. Both programs are dangerous and have several inherent risks if certain precautions and protocols are not taken. Many shark diving encounters are based on volunteer standards and recommendations, but a few are government-regulated. For example, shark diving in South Africa is regulated by the South African Department of Labor. In Florida in 2001, the FFWCC adopted a law preventing the feeding and baiting of sharks.

Boating and SCUBA Accidents, Low-Head Dams, and Beach Renourishment   273

FIGURE 14.4   Shark encounters are becoming more popular, but not without risks. "Chumming" and feeding is a practice that invites attacks and should never be practiced.

While shark attacks are relatively rare events, under certain circumstances the chances of being attacked increase dramatically. These circumstances include the following:

- Many species of sharks are crepuscular in their feeding habits, meaning that they feed during early morning and evening hours so conducting encounters during these times when visibility is low is dangerous. Poor visibility prevents sharks from distinguishing prey species (i.e., seals, turtles, and fish) with humans. Research has shown that when visibility is low, sharks are more reliant on their sense of taste rather than their sense of sight.
- Conducting encounters in areas where particularly dangerous shark species, including the shortfin mako shark, great white shark, dusky shark, and bull shark, are known to exist. Whenever a dangerous shark is observed, divers should immediately be evacuated from the water. Alternatively and preferably, shark cages should be used.
- Conducting encounters where baiting is used to attract sharks. Sharks are attracted to blood in the water and they can detect blood in low concentrations. Many dive boats use "chum" or bait to attract sharks deliberately for their customers. Chumming may initiate "frenzy" feeding behavior. During frantic frenzy behavior, sharks often mistake humans for their prey species. In many areas, shark feeding is illegal and in some cases, despite being illegal, shark dive operators ignore laws prohibiting feeding and continue this dangerous practice.
- Conducting encounters in large groups. In the majority of cases, sharks will be attracted to a large group of divers as opposed to a smaller group.

The Hurgada Environmental Protection and Conservation Association (HEPCA) recommends the following safety rules when swimming with sharks:

- No swimming or snorkeling in waters where large shark species are known to exist.
- No deliberate feeding of sharks or dumping of waste from boats which may attract potentially dangerous sharks (these activities are illegal in many areas of the world).
- No SCUBA diving without an experienced dive guide in waters where large shark species are known to frequent.
- In areas where sharks frequent, divers should enter the water as quietly as possible and without creating a disturbance.
- Never dive at night or during dawn or dusk.

In any shark encounter program, it is incumbent for the operator to conduct a site inspection to determine the species of sharks, their relative abundance, and level of aggressive behavior being exhibited by the local shark population. Many shark populations have been habituated to humans and usually do not demonstrate aggressive behavior, while other populations (mostly pelagic species) may be more aggressive when encountering and interacting with humans.

In addition, shark encounter operators must have a carefully designed and articulated risk management plan. This includes rehearsing a plan focused on treating a shark bite victim. Every member of a shark encounter team should have an advanced and current emergency first-aid certification. In addition, every member of the team should be professionally trained and certified by a recognized diver training program, for example, NAUI or PADI. Finally, every shark encounter guide should have considerable background and experience in dealing and interacting with sharks. An inexperienced guide may not be able to detect and respond to the many risk factors associated with shark diving. Ultimately, any injury sustained by a customer is the responsibility of the guide and the organization promoting the encounter. Because sharks are so unpredictable, there is considerable liability to any educational institution, organization, or company, either directly or indirectly involved in this activity.

LOW-HEAD DAMS

Low-head dams are sometimes called "drowning machines." There is a good reason why they have been called this—over the past 40 years, low-head dams have been responsible for more than 100 fatal drownings and several thousand serious accidents requiring emergency first aid intervention (see Figure 14.5).

A low-head dam is a low-profile dam, ranging between 5 and 12 feet high. While estimates vary, there may be as many as 80,000 low-head dams in the United States, and they are found in every state. See Figure 14.6. Most low-head dams were constructed a long time ago, some at the turn of the last century. Many low-head dams have been abandoned and are in a terrible state of disrepair. Low-head dams sometimes create barriers, preventing fish from migrating further upstream, and are responsible for creating several other significant environmental impacts. Consequently

Boating and SCUBA Accidents, Low-Head Dams, and Beach Renourishment 275

FIGURE 14.5 Low-head dams are often responsible for creating hydraulic currents, which have been responsible for more than 100 drownings.

FIGURE 14.6 There are an estimated 80,000 low-head dams located in the United States. Here is example of a low-head dam where fatal drownings occurred.

responding to these problems, many low-head dams have been breached rather than removed. Breaching is usually much less expensive than removing, so this is a much more common practice even though breaching may create a danger to the public.

Most low-head dams were built as power sources, taking advantage of a river or steam's strong current. At the turn of the century, many low-head dams provided the power to operate grist mills. Some low-head dams served a dual purpose. Sometimes, this second purpose was to store water for irrigation, especially during times of drought. A few low-head dams continue to serve in this capacity.

Low-head dams have been built using several methods and materials. There are six types of low-head dams, including concrete gravity dams, concrete arch dams, concrete buttress dams, earth dams, earth and rock dams, and concrete-faced rock fill dams.

All low-head dams represent significant safety hazards, primarily to individuals engaged in recreation. Canoeing, kayaking, rafting, fishing, and bathing represent activities that commonly occur at or near low-head dams. In many cases, the sudden drop created by many low-head dams creates a "thrill" that individuals engaged in paddle sports often find hard to resist.

Low-head dams create a public safety hazard because of the fact that these structures often look harmless or sometimes even inviting to recreational users. The danger of these structures is that the downstream side contains a submerged hydraulic jump or "hydraulic." In the most basic terms, a hydraulic is a strong, circulating current that extends from the surface to the river bottom. Depending on the height of the dam, the depth of the water, the width of the river, the slope of the embankment, and the current velocity, the strength of the hydraulic may vary significantly.

In most cases, when caught in a hydraulic, a victim finds it difficult and often impossible to escape. A typical scenario involves a victim who is trapped and continuously circulated in the current until he or she becomes fatigued and drowns. Escaping a hydraulic is exceeding difficult and often impossible. In a lucky few cases, and if a good swimmer, a victim, upon reaching the river bottom can kick off from the bottom with enough power and momentum to escape downstream.

> The hydraulic jump creates a recalculating current which can trap watergoers in a seemingly endless cycle of being pulled under, struggling back to the surface, being pushed back toward the falling water, and once again being pushed underwater. (Elverum and Smalley, 2003)

The best known and most publicized low-head drowning event was in Binghamton, New York, in 1975 when three people drowned (two canoers and one rescuer) and four other rescuers were seriously injured. This tragic event was captured on camera and demonstrated to the public the serious threat that low-head dams present to the public.

Another highly publicized low-head dam multiple drowning event occurred in Minnesota. In July 1977, a twenty-five-year-old man went over a low-head dam while floating on an air mattress. He immediately became trapped in the hydraulic current and was not able to escape. Two people in canoes attempted to help but soon they both became trapped in the deadly current. A state trooper and several other bystanders attempted to rescue the three victims, but failed. Soon thereafter, all three victims drowned.

Establishing ownership and safety of low-head dams is often problematic because several agencies may have jurisdictional responsibility. In the majority of cases, a government entity has the ultimate responsibility for public safety. However, in some cases, this may be shared with a private entity such as a rafting or canoeing trip company that derive income from using the river.

As with any hazard threatening the lives of the public, at the very least, signs should be placed well upstream of low-head dams warning the public to stay clear and about the possible consequences of paddling or swimming over a dam

Boating and SCUBA Accidents, Low-Head Dams, and Beach Renourishment 277

FIGURE 14.7 Because low-head dams represent a proven safety hazard, warning signs must be posted to bathers, kayakers, rafters, and canoers.

(see Figure 14.7). In addition, warning signs should be placed on either side of the dam, both above and below. Signs should adhere to ANSI warning sign standards and should be regularly maintained. If damaged, vandalized, or faded, they should be replaced. If possible, emergency ring buoys should be strategically located within reach of the public.

Public education is also important when preventing low-head drowning events. Education signs can be effective if strategically placed in the vicinity of low-head dams. Signs should describe a "hydraulic" and should include instructions on how to escape a "drowning machine."

When a low-head dam is located within their jurisdiction, all public safety personnel (police, fire, and paramedics) should receive professional training focusing on low-head dam and swift water rescue methods. Low-head dam rescue requires considerable technical knowhow, and in the absence of this training, there is the strong likelihood of a multiple drowning event. Consider the two drowning events referenced in the above.

In addition to training, all police and fire departments should have the equipment necessary needed to perform effectively a low-head dam rescue, including ropes, anchors, and at least one inflatable rescue boat (IRB). In some cases and providing that there is good access, rescues can be performed using a hook-and-ladder truck by extending the ladder over the victim and lowering a safety line with a buoy attached.

In rare cases, a helicopter may play a key role in making a low-head dam rescue. However, since response time is a critical factor, there is usually not enough time for a helicopter to become airborne to arrive on the scene in time to affect a rescue.

BEACH RENOURISHMENT

In a preliminary investigation, it was discovered that beach sand nourishment may have consequential and serious impact on some beaches. Over the last several years, significant increases in fatal drowning and injuries have been reported on beaches that have been recently nourished (Fletemeyer et al., 2017). On a mid-Atlantic beach in Maryland, a 397 percent increase in serious accidents was reported following a beach sand nourishment program (Chesler, 2013). The reason for this is related to the fact that when beaches are nourished, their slopes and compaction values are drastically modified. High-profile beach escarpments often form along the water's edge. These conditions may result in wave runback at increased water volume and velocity, especially during a moderate and heavy surf environment. In turn, this may result in more and stronger rip currents and the formation of plunging and spilling waves that are responsible for spinal injuries.

Due to a rise in sea level, many beaches are suffering from sand erosion. This seriously diminishes a beach's recreational value. Consequently, many beaches are being renourished with sand. Sand for renourished beaches comes from several sources and is deposited in three ways:

- Sand is hydraulically dredged from offshore sand deposits.
- Sand is excavated from offshore deposits and then barged to the beach for offloading.
- Sand is dump-trucked from offsite land sources.

Once deposited on the beach, the sand is spread with large graders and backhoes to specific engineering specifications. Beach renourishment is never permanent because erosion is a continuous process. In most cases, renourishment is designed to last six to eight years, requiring a follow-up program. In some cases, when a significant storm event impacts the beach, such as a hurricane, a beach might revert back to its original erosion state in only a day; hence, a project costing several million dollars may be lost and an emergency program is needed.

Beach renourishment remains controversial because of the cost and because it disrupts a naturally occurring process. In addition, renourishment impacts several species of flora and fauna that inhabit the coastal zone, including marine turtles and sea birds.

In addition, there is evidence that beach renourishment impacts public safety. The United States Lifesaving Association reports that serious injuries increase threefold on beaches that have been the recent subject of renourishment. There is anecdotal information suggesting that beach renourishment is responsible for increasing the number of spinal injuries.

In 2014, at the International Pip Current Symposium in South Korea, Fletemeyer reported that beach renourishment may be responsible for increasing the strength of rip currents. This was based on data indicating that renourishment programs are responsible for creating a significantly more compact beach. For example, on some renourished beaches, due to the soft sand, driving a vehicle is difficult and sometimes not possible. Post-nourishment, the beach becomes more compacted, allowing for driving vehicles on the beach. The distribution of sediment and the shape of sand grains was suggested as the reason for the difference in compaction. A microscopic examination of sand revealed that sand grains on natural beaches are more round and weathered than sand deposited on renourished beaches. This observation was made on a selected number of Florida beaches.

As a result, many beaches as a condition of permitting must initiate a sand-tilling program that functions to soften the beach and allows it to revert to its original state. It has been hypothesized that a more compact beach prevents water from percolating through the sand, causing more water to wash out to sea and consequently increasing the strength and velocity of rip currents. In addition, there is suggestive evidence that beach renourishment may be responsible for influencing wave dynamics and the creation of nearshore bars.

On land, beach renourishment is often responsible for the creation of beach scarps. On some renourished beaches, scarps may be nearly vertical and several feet in height. Scarps of this size create a hazard to pedestrian traffic, especially at night.

REFERENCES

Chesler, C. 2013. Do serious injuries come in waves? *Scientific American*, June (4). pp. 11–14.

Elverum, K., and Smalley, T. 2003. The drowning machine. Boat and Water Safety Section, Minnesota Department of Natural Resources.

Fletemeyer, J., Haus, B., Hearin, J., and Sullivan, A. 2017. The impact of beach sand nourishment on beach safety. *Journal of Coastal Research*.

15 Hydrodynamic Analysis of Drowning Events

Brian K. Haus

CONTENTS

Introductory Conversation 281
Initial Overview 282
Summary 289
References 289

There is a distinct difference in the way that scientists present their conclusions about a subject of interest between communications targeted to their colleagues and those to stakeholders outside the scientific community. This is particularly true in legal proceedings where simple definitive statements are required. Fortunately, however the process of analyzing a particular problem is very similar in both situations. In this chapter I will describe the steps taken in determining the role that physical processes including waves, winds, currents, and tides may have played in an event that is under litigation. I will break these up into four sections divided along a timeline that positively correlates to commitment to the case (financially and intellectually) and in the knowledge of the circumstances involved.

INTRODUCTORY CONVERSATION

It has been my experience that there are often existing misconceptions about hydrodynamic processes and their potential role in a water-related accident that need to be addressed before proceeding. The introductory conversation should address the following key questions:

1. Is this something within my area of expertise?
    Upon being contacted by one of the parties in a suit, it is necessary to first determine if the questions of interest to the person initiating the contact are within the boundaries of my professional expertise. While this may seem at first to be a relatively simple thing to determine, it is often the case that the lawyer involved does not have a clear idea of what the scope of a physical oceanographer/coastal engineer's expertise is. Additionally, as an active researcher I publish papers on much more detailed aspects of particular processes. Therefore, the most important thing to do is to have a conversation about what aspect of the case the lawyer is interested in engaging your expert opinions on and what the general scope of the issue is. Once this is clear and

it is mutually agreed that I likely have the appropriate expertise to contribute to the case, it is time to proceed to the basic facts of the case.
2. What are the agreed-upon facts?

The party that has contacted you will certainly be presenting a version of the case that is colored by their own interpretation of events, but it is important to establish what is known and agreed upon by both parties in the dispute. That allows you to focus on the questions that are pertinent to your participation in the course. For example, it may be understood by both parties that a victim drowned at the beach, but the precipitating factors may not be understood.

3. What is under dispute, and how am I expected to contribute to the case?

There are a number of circumstances in which the ocean water and its motion in the form of waves, currents, or tides (hydrodynamics) and related wind (meteorological) factors are not in dispute, or neither the defendant nor plaintiff have asserted that they are a contributing factor to the case. In this situation it is unlikely that an oceanographer/coastal engineer such as myself will be contacted, but preliminary review may show one of three things: First, that the facts of the case that are under dispute do not fall within my area of expertise and other experts should be contacted; second, that although the present scope of the case does not include issues that directly call upon my expertise, the hydrodynamic conditions may have posed issues that should/could be raised by one of the parties and my contributions may be needed at some point. The third result of initial discussions is the realization that the case does indeed include contested issues related to hydrodynamics and that my expertise could shed light on them. In this case, an extended conversation is conducted in which I outline to the legal representative how I believe my expertise could contribute to the case and what the range of findings could be.

4. What are the logistics?

At this stage it is necessary to discuss and agree upon an initial scope of work required to reach a preliminary expert opinion on the hydrodynamic conditions pertinent to the case. Here, there is a situation where I as the expert am trying to provide sufficient information for the lawyer to determine whether I should be retained, while limiting time invested up front. As a scientist I would refer to this as developing a hypothesis. This term is widely misunderstood in the general public as a guess; here I use it in the way a scientist would understand it. It is what, upon this introduction to the case and weighing the limited available evidence at this stage, I believe the most likely cause is. Further investigation should be focused on testing the validity of this hypothesis with increasing degrees of resources.

INITIAL OVERVIEW

An initial overview of an incident usually involves reviewing available documentation provided by the attorney with whom the initial conversation took place. Here, official reports that document the time and location of the event are particularly useful. If such documents are not available, then a first review of eyewitness testimony (often obtained from depositions) is required. Later in-depth review should follow, but initially the following key points should be addressed.

1. Is the location of the incident known? If so, what is the physical setting?

    The basic question of where an incident occurred can greatly determine what course an investigation will take. If this basic piece of information is unknown or under dispute, the analysis process is considerably more time-consuming. Determining the likely location will require applying what is determined in the hydrodynamic analysis to either project forward in time from a last known position or backward in time given a location where an injured or deceased person was found.

    - If official reports and eyewitness testimony establish location of incident.

        This is the single most valuable piece of information because most coastal regions, rivers, or lakes experience a limited range of waves, currents, or tides. Although this can certainly vary seasonally and with extreme events such as hurricanes, it is reasonable to expect that there is some predictability to the ranges over which they vary in general. Here one must be careful, however, to understand that small-scale variability may play an important role in a particular case. Rip currents are the most obvious example of this type of variability. On a long, open beach the average conditions may not appear to be hazardous, but local strong rip currents may cause an extreme hazard. The challenge is to identify the types of mean predictable conditions that might produce these strong local events. The meteorological parallel is tornado forecasting, in which it is impossible to predict the exact time and place a tornado will form, but conditions that lead to their formation are known and can be identified. Here, working backward from some event, one is tasked with attribution of cause for some event which requires understanding the mean local conditions and what variations from that mean might produce elevated hazards. Knowing where something occurred is the critical first step in that process.

    - If incident occurred on or adjacent to land.

        I expect that it is no surprise when I say that the next step after learning the location of an incident is to use Google Earth to view the location. This resource represents one of the greatest, if not *the* greatest, contributions to coastal sciences, and it is getting more powerful by the day as imagery is added to the archive. It is important to look at the site over a range of scales, from the smallest scale that the imagery can support (to see details of local topography) to the much larger view for information on potential wave, tide, and current effects.

    - If incident occurred in open water.

        If the incident occurred in a location without a specific geographic reference, then the issue of location can become significantly more challenging. There are many examples of such incidents, including a victim swimming, skiing, diving, or tubing off a boat. While GPS positioning is widely available today from chart plotters, handheld GPS, or smartphones, it is not unlikely that persons involved in a maritime incident will neglect to record a position. In such a case, approximate positions are needed to establish the general area of the event (i.e., in a bay, inlet, shallow shelf water, or open ocean), because the hydrodynamic conditions may be very different in each case.

In the most difficult situations there are no eyewitnesses and limited knowledge of where the incident occurred. An example could be when a person boards a rental Jet Ski, boat, kayak, paddleboard, etc. without an on-board GPS and then leaves the immediate area where he or she is under observation. Then, sometime later, his or her body (either on or off the rental equipment) is discovered floating in another location. Depending on the complexity of the local marine, river, or lake area, determining the location of the event can depend on a lengthening chain of assumptions and may require significant effort in flow hindcasting to establish with any certainty the conditions that led to the event.

- Trajectory analysis.

The science of doing backward trajectory analysis requires implementation of a computer model of water circulation for the region of the incident. This is not a simple undertaking because numerical circulation models of particular bodies of water require substantial input data and calibration and validation before they can be trusted to provide accurate transport estimates. If no calibrated model exists for the region of interest, it is a time-consuming and expensive endeavor to implement one. It is not simply the computer costs associated with running the model, but rather the cost in experimentation to provide appropriate initial fields (topography, currents, winds) and boundary conditions (flows entering or leaving the domain of interest) and to calibrate and validate the output waves or currents. Collecting such information is expensive and often beyond the resources that a law firm is willing to invest in a case. While model simplifications can be invoked to reduce the expense, any simplifications must themselves be tested.

2. How well is the time of the incident known?

While the answer to this question is usually correlated with the answer to the previous question, there can be cases where the location is well established but the timing is not known within a wide range from weeks, days, hours, and so on. Usually this is the case of a body being discovered at some location that does not have the capacity for significant transport, such as a small lake or spring. Another situation that might occur is that the decedent is pinned or entangled in such a way that they could not have moved from the location where they were found. The timing of the event that led to the fatality can then be of critical importance for determining the human-related aspects of the case, that is, visibility of hazards, fatigue, sun exposure, and ancillary risks such as alcohol or drug consumption. The timing is also important for providing an initial constraint on how well the hydrodynamic conditions can be determined.

Determining the time at which the event occurred that led to a death is primarily an undertaking that is outside the scope of the oceanographer's expertise. It relies on determination of a timeline based on an autopsy, physical evidence, and witness testimony. There may be a role in providing corroborating physical evidence to establish when it would be possible for particular situations to occur. As an example: At what times of the day could the water level have been high enough for a swimmer to enter a particular beach? Here, the timing and the assessment stage that will be discussed next become intertwined because the hydrodynamics must be used to help establish the key question of when.

3. Hydrodynamic conditions initial assessment.

The initial hydrodynamic forcing should be investigated to gain an idea of the relative magnitude of the importance of waves, currents, tides, and wind in an incident. While all of the potential contributing factors need to be addressed before reaching a conclusion, as a starting point the most likely significant forcing should be determined. This will depend strongly on the location and time of the incident, as discussed previously.

As an example: If the incident occurred in sheltered waters such as an estuary or inlet, the direct wind and wave forcing are likely to not be the most important factors. However, tidal forcing can be strong in these areas because the flow is constrained by topography.

The opposite condition is on a long, open coast exposed to significant wave energy. Here the direct wave action is likely important, but so are the waves driven alongshore currents and rip currents. These currents are generated by the energy imparted by breaking waves, but may be influenced to some degree by local wind and tide conditions. Large-scale breaking waves carry energy built up from the wind forcing over ocean basins and deposit it in the very narrow strip along the beach we call the *surf zone*. Some of this energy lost by the wave field drives mean longshore currents when the waves approach at an angle to the beach.

- Long-term conditions.

    To determine the likely impact of wave action on the beach in question, one must consider the regional wind–wave climate, the local topography, and the conditions on the day of the event. The wind–wave (waves that are forced directly by the wind) climate can vary significantly seasonally in some areas. For example, the Gulf of Mexico in the summer is often quite calm, with very low wave energy. In the winter it is strongly impacted by passing weather systems, causing the wave energy to vary on scales of days to weeks. In regions exposed to the open ocean, distant "swell" waves (that is, those that are not forced directly by the local winds) generate large breaking waves and nearshore currents.

    The local topography can determine if the offshore waves present on a particular day are likely to propagate into the beach. If there are offshore barrier islands or island chains (Bahamas in South Florida, Catalina Islands in Southern California) they can significantly impact the way in which incoming waves approach a local beach. For example, the Bahamas serve to reduce the *fetch*, the oceanographic term for the distance in which the wind blows over water and generates waves, in South Florida to that approximating the width of Lake Michigan. This substantially reduces the potential size of waves approaching South Florida beaches. It also causes the dissipation of most "swell" waves, limiting potential approach angles from the open ocean. If the offshore depths on a particular beach are shallow enough to break incoming waves before they reach the beach in question, this will also shelter it from incoming waves. The amount of sheltering of this type will typically vary with tidal levels. Low tide provides for more protection than high tide because of the shallower water over the offshore shoals.

- Daily conditions.

To develop a picture of what the conditions were on the day of the event, there are essentially three sources of information. The first is any eyewitness testimony regarding the winds, waves, and currents at the time and location of interest. The second is from data mining: using meteorological or oceanographic data or available model hindcasts to find information in close proximity to the event. The third approach is to develop a model hindcast for the local conditions during the event. The first two sources are available at this initial stage of review, while the effort required to develop, validate, run, and interpret a computer model is beyond the scope of an initial investigation. This approach will be discussed at the end of the chapter.

While eyewitness testimony is typically not available in quantitative terms, it can provide a local, timely, qualitative description of the conditions. Key things to be aware of are any testimony regarding the local wave heights and direction of approach. Were the waves breaking? If so, was the height of the breaking waves estimated? Was there a longshore current at the beach? How strong were the winds that day?

The next step is to look for the nearest recorded wind, wave, or current or tide level data to the location of the incident. The best place to start for a coastal event is at the U.S. National Data Buoy Center (NDBC; see http://www.ndbc.noaa.gov/rmd.shtml, station locator page). This page is the entry point for finding stations that report data to the NDBC; it also covers any data available anywhere in the world. However, in some regions the available data is quite limited. To be included in this data stream the measurements must meet NDBC standards for data quality, metadata, and archiving, which is useful for inclusion in court proceedings. Meteorological measurements are quite common in this database (wind speed, direction, air temperature), with wave measurements only available at a subset of the locations and ocean current observations being yet more sparsely located.

From these disparate sources of information, it may be possible to form a reasonable opinion on the hydrodynamic conditions present at the time and location of an event of interest. The constraints on application of the nearby data are the distance from the measurement to the location of interest, the latency of the data (how much time separation from the observation to the incident), the steadiness of the forcing in both space and time, and the complexity of the local coastlines.

The relative importance of the distance in space and time separation are related to the steadiness of the forcing (winds) in each. If a large atmospheric circulation pattern is set up, such as the North Atlantic high, the wind fields generated by this are uniform over hundreds of kilometers and a quite distant measurement can be reflective of local conditions. This contrasts with summertime convective activity in the Southeast United States, where thunderstorms with spatial scales of tens of kilometers and temporal scales of minutes to hours may be important. In this case the data must be much closer in space and time to the event of interest to be relevant for the local conditions.

4. In-depth analysis of hydrodynamic conditions.

There is a narrower range of cases where the hydrodynamic conditions are critical to the outcome and the resources are provided to do a thorough modeling study. Here I will discuss the steps required to implement such a study. The first being to select the models of interest based on the location and the expected relative importance of wind, waves, currents, and tides. The second step is to configure the model including the topography to be used. The initial conditions and boundary conditions to be used must be determined. At this stage the model must be validated, preferably with data at the site of interest.

- Model selection.

    There are an ever-growing set of wave and circulation models available for nearshore hydrodynamics. In practical terms, the choice of which to use in a particular study is driven in large part by the models you have experience with, because of the large amount of time required to understand appropriate implementation. So, in effect, the choice of model to use is made before there is a specific application of interest. In order for a model to be useful for a wide range of potential marine incidents it must have the following characteristics: It must have the ability to implement complex topography; it must be capable of producing both wave and current estimates or be able to be coupled with models that can do so; it must be computationally inexpensive so that a range of scenarios can be run without requiring very long run times or using supercomputing centers; and finally it must be widely used so that its results are well established in the peer-reviewed literature. An example of such a model is the Delft3D circulation model, which is an open-source community model and as such has been widely used and tested. It is usually coupled with the Simulating Waves in the Nearshore (SWAN) wave model to provide wave driven current components. An in-house research model may work very well for particular problems, but if it isn't established scientifically then its results will be subject to challenge in court.

- Model configuration and collection of topography.

To run a nearshore wave and circulation model, it is critical to set up the model domain large enough to incorporate the likely scales of motion and to have fine-enough grid resolution to resolve the key topographic features. The optimal trade-off between model size and grid resolution is constrained by the computational expense. It is also important to be able to define the topography of the model in the region of interest and to have a strategy for realistically forcing the model with tides, winds, or larger-scale currents, depending on the relative importance of these processes and the availability of appropriate data from local measurements or from established larger scale models.

Accurate, high-spatial resolution topography that was collected recently enough to be representative of the time when the event of interest occurred is often unavailable. This represents one of the biggest challenges when attempting to hindcast hydrodynamic conditions at a particular location. This is particularly challenging in sandy areas with significant sediment transport. The locations of sand bars on open coasts and spits and sandy shoals in inlets and estuaries can shift considerably in a single storm event. Therefore, even if the topography is known very well at the current time because a survey was conducted, it cannot be immediately assumed that this is the topography present at the time of the event. A review of previous shoreline positions using Google Earth, for example, can reveal something about how much variability in the shoreline position has occurred at the location.

- Specification of initial and boundary conditions.

The best way to provide initial and boundary conditions for a nearshore model is to have them provided by a larger-scale operational model such as the Naval Research Laboratory's Real-Time 1/12° Global HYCOM Nowcast/Forecast System (Jacobs et al., 2014). This model uses available satellite and in situ data through the process of data assimilation to produce the best available ocean state forecasts and nowcasts for all of the world's oceans. It can then provide a coarser resolution initial state as well as flows at the offshore boundary of the model. If larger-scale currents are not likely to impact the local domain, it is acceptable to force the nearshore model with an offshore tidal variation, winds, and waves. This greatly simplifies the implementation of the model.

- Model validation.

The time spent configuring a nearshore model to study a particular region is significant, but it pales in comparison to the effort required to validate the model for the hydrodynamic processes of interest. It is this step that typically constrains the use of these advanced techniques for legal cases. Validation requires that realistic forcing be applied to the model and that the simulated response of the nearshore region be compared with measurements in the domain of interest. These measurements can be based on moorings or data collected from moving platforms or drifting sensors. The data collection must be designed to capture the key processes that are likely to be important for the case under consideration. Rip currents in particular are difficult to sample because of their intermittency in space and time and the fact that they occur in the highly energetic surf zone. Here it is usually necessary to have a combination of bottom-mounted moorings to capture the currents and waves for a significant length of time (over weeks) and then combine this with drifter studies (e.g., Brown et al., 2015) to identify the spatial variability of the rip currents.

SUMMARY

The process of evaluating the hydrodynamic contribution to a drowning or personal injury event is comparable to a scientific research project. The stages of inquiry can be posed in similar terms, because it is about establishing a hypothesis and testing with the available information. The amount and type of information available is highly variable in regards to location and timing. If resources are substantial, then available information can be supplemented with additional observations and modeling that can shed significant light on the important processes.

REFERENCES

Brown, J. A., MacMahan, J., Reniers, A. J. H. M., and E. B. Thornton. 2015. Field observation of surf zone-inner shelf exchange on a rip-channeled beach. *Journal of Physical Oceanography* 45(9), 2339–2355.

Jacobs, G. A., Bartels, B. P., Bogucki, D. J., Beron-Vera, F. J., Chen, S. S., Coelho, E. F., and Curcic M. et al. 2014. Data assimilation considerations for improved ocean predictability during the Gulf of Mexico Grand Lagrangian Deployment (GLAD). *Ocean Modelling*, 83, 98–117.

Index

Note: Page numbers followed by "f" indicate figures.

A

Aboveground pools, diving accidents, 114–116; *see also* Diving accidents
ACA, *see* American Canoe Association (ACA)
Accidental deaths, 63
Accidental injury, vii
Accidents; *see also* Injury(ies)
    diving, *see* Diving accidents
    drowning, *see* Drowning
    psychological evaluation, 237–247; *see also* Grief/grieving process
    SCUBA, 265–274; *see also* SCUBA
    whitewater rafting, 189–191
        death causes, 189–190
        fatalities, 189
        injuries, 190–191
Acidosis, 38
Active drowning, 31–32, 35
    factors contributing to, 36f
Adika v. Beekman Towers, Inc., 199
Advanced (Class IV) rapids, 175
Air embolism, 269
Alcohol
    boating under the influence of, 263–264
    demonstrated effects of, 264
    hot tub drowning, 79
Ameba, *see Naegleria fowleri*
American Canoe Association (ACA), 182
American Lifeguard Association, 46–47
American Red Cross, 28, 47, 139
American Whitewater, 175
Anti-entrapment drains, 58, 81–82
Apartment pools, operation and management of, 134
Apnea
    initial, 34
    terminal, 34–35
Aquatic law, *see* Law
Aquatic Risk Management, 58
Archimedes principle, 37
Ard, Anthony, xi
Army Corps, 158
Assertiveness training, 244
Assumption of the risk, 6
Attractive nuisance doctrine, 14–15
Auer, A., 255
Automated external defibrillator (AED), 38, 183–184
Autonomic nervous system, 108, 109f

Autopsy, 63
Avramidis, S., x

B

Barotrauma, 269
Barrier fences, swimming pools, 85–87
Bathing, lakes for, 158–160
Bathing load, swimming pools, 92–93
Beaches, 138–139
    case study of drowning, 61
    colored flags, 52, 54, 54f
    materials composing, 138
    negligence considerations, 140
    population visiting, 138, 139f
Beach hazards, 140–157
    dangerous activities, 146–148
    docks, 143–144
    dogs, 148
    driving, 145–146
    escarpments, 149
    fires, 148
    groins, breakwaters, and jetties, 142–143
    jellyfish, 156
    lighting, 154
    lion fish, 156
    outfalls, 144–145
    piers, 141–142
    Portuguese man-o-war, 156
    rip currents, 150–152
    sea cliffs, 157
    shallow water diving, 152–153
    sharks, 155–156
    stingrays, 156
    submerged rocks and boulders, 143
    water pollution, 154–155
Beach recreation, 138
Beach renourishment, 278–279
Becker, R., 250
Behavioral effects, of grief and traumatic injuries, 242
Behaviors, drowning
    in active drowning, 31–32
    perception commonly held by the public about, 37f
Bereavement, 242; *see also* Grief/grieving process
Biological death *vs.* clinical death, 30

The Biological Evidence Preservation Handbook: Best Practices for Evidence Handlers, 64
Blacka v. James, 10–11
Bloated stage, of body decomposition, 251
Boating accidents, x–xi, 262–263
　alcohol, 263–264
　case study, 259
　inexperienced boaters, 265
　operator error, 262
　scenarios defining, 262
Boating under the influence (BUI), 263–265
Boat order, whitewater rafting, 187–188
Boats, safety equipment, 263
Body decomposition, stages and description of, 250–252
　bloated stage, 251
　decay stage, 251
　fresh stage, 250
　skeletal stage, 251
Bradycardia, 39
Breach of duty, 27
Breakwaters, *see* Groins, breakwaters, and jetties
Breaux, Frederica, 197
Breaux, Zachary, 197
Breaux v. City of Miami Beach, 197–208
Bureau of Land Management, 180

C

CALEA, *see* Commission on Accreditation for Law Enforcement Agencies (CALEA)
Calhoun, Natalie, 213
California, beaches in, 138
Camargo, Ruben, 105–108, 110
Canals, 164–165
Captain (boat operator), 262–263
　responsibility, 263
Cardiac arrest, 35
Cardio pulmonary resuscitation (CPR), xiv–xvi, 38
　certification, 182
Carter, R. F., 224
Carter v. Boys' Club of Greater Kansas City, 13
Case studies
　drownings/accidents and risk management, 61–62
　Ft. Lauderdale new river drownings, 57
　large dam boat accident, 259
　low-head dam drowning, 258
　shallow water diving accident, 258
　swimming pool drowning, 259
Case Studies in Drowning Forensics (Gannon and Gilbertson), 252
CBT, *see* Cognitive behavioral therapy (CBT)

CDC, *see* Centers for Disease Control and Prevention (CDC)
Centers for Disease Control and Prevention (CDC), x
　on disinfection of water, 94
Certification program
　river guides, 182
　SCUBA, 265–266
Chemistry, swimming pool, 93–94
Chick Kam Koo v. Exxon Corp., 218
Child-proof barriers, 80
Children
　parental concerns/note, 135–136
　private pools and, 76
　swimming pool drowning, 80
Chlorination, for *Naegleria fowleri*, 228
Clinical death, 38
　vs. biological death, 30
Coan v. New Hampshire Dept. of Environmental Services, 7–8
Coastline of the United States, 138
Coblentz Agreement, 205, 206
Cognitive behavioral therapy (CBT), 244–245
　assertiveness training, 244
　interoceptive exposure, 245
　mindfulness-based approaches, 244
　PCT, 244
Cold water, 39, 179
Coleman v. Shaw, 9
Cole v. South Carolina Electric and Gas, Inc., 8
Color-coded pool edges, 131
Colored flags, as signs, 52, 54, 54f
Commercial whitewater rafting, *see* Whitewater rafting
Commission on Accreditation for Law Enforcement Agencies (CALEA), 65
Comparative negligence, 6
Condo pools, operation and management of, 134
Consent, 2
Construction-related drowning, 41
Consumer Products Safety Commission (CPSC), 74
Contributory negligence, 6, 26–27
Corda v. Brook Valley Enterprises, Inc., 15–16
Cost of drowning, 29–30
　economic damages, 29–30
　non-economic damages, 30
　punitive damages, 30
Court of Appeal of Louisiana, 3
Cousteau, Jacques, 265
CPR, *see* Cardio pulmonary resuscitation (CPR)
CPSC, *see* Consumer Products Safety Commission (CPSC)
Cracraft v. City of St. Louis Park, 229–230
Crypto (*Cryptosporidium*), 93
Cyst form of *Naegleria fowleri*, 227f, 228

Index

D

Daily conditions, hydrodynamic analysis, 286–287
Damages, 28
DAN, *see* Divers Alert Network (DAN)
Dangerous activities, beaches and, 146–148
Daubert standard, 69–71
Daubert v. Merrell Dow Pharmaceuticals, Inc., 69, 70
Davison, Michelle, 100
Death on the High Seas Act (DOHSA), 209–219
    congressional debate on, 216
    enactment (1903-1920), 215–219
    fixing, 218–219
    geographic application, 219
    high seas, 217–218
    Mann Amendment to, 216, 218–219
    wrongful death claims prior to, 214–215
Deaths
    accidental, 63
    biological *vs.* clinical, 30
    events surrounding, 243
    from free-living amebae, *see Naegleria fowleri*
    in hot tub, 79
Decay stage, of body decomposition, 251
Decks, swimming pools, 84
Decompression sickness, 268–269
Deepwater Horizon, 213–214
Delft3D circulation model, 287
Depecage, 213
Depth awareness, 130
Depth marker, swimming pools, 90, 129–130
    florescent, solar or neon nighttime, 130
The Diagnostic and Statistical Manual of Mental Disorders, fifth edition (*DSM-V*), 237
Diarrheal illnesses, 93
Diatoms, 254, 255f
Distress, 30–32
Divers Alert Network (DAN), 265
Dive training organizations, 266
Diving
    competitive, 100–101
    hazards, 267
    lessons, 101–102
    sport of, 100–101, 102
    usage of word, 102–103
Diving accidents, 100–136
    aboveground pools, 114–116
    brief history, 100–102
    categories, 112
    data and facts, 112–114
    examples of, 103–108
    inground pools, 116–118
    insurance and medical industry, 134–135
    natural environment, 123–125
    parental concerns/note, 135–136
    reducing, 127–135
    reenactments, 256–258
    shallow water, 125–127
    springboard/jumpboard, 118–121
    swimming starting blocks, 121–123
Diving boards, swimming pools, 85
Diving Injuries (Gabrielsen), 102, 107, 112, 133
Diving reenactment, 257
Docks, as beach hazard, 143–144
Dogs, as beach hazard, 148
Dontas v. City of New York, 8–9
Drowning, vii
    active, 31–32
    case studies, 61–62
    causes of, 28–29
    cold water, 39
    construction-related, 41
    cost of, 29–30
    definition of, 28
    economic cost of, xvi–xvii
    entrapment, 41
    hydrodynamic analysis, 281–289
    as a leading cause of death, 1, 28
    number by age groups, 35, 36f
    passive/silent drowning, 38
    physiology, 33–38
    psychological evaluation, 237–247; *see also* Grief/grieving process
    reenactments, 256–258
    salt *vs.* fresh water, 32–33
    toddlers, 40
    work-related, 41
Drowning machines, *see* Low-head dams
Drugs, hot tub drowning, 79
Dry drowning, 2, 15–16, 33
Duties of governmental owners, 228–232
Duty, 27
    breach of, 27
Dyspnea, 34

E

Easy (Class I) rapids, 175
Economic cost of drowning, xvi–xvii
Eddy, 176–177
Eddy lines, 176–177
Electrocution, as hot tub and spa hazard, 79
EMDR, *see* Eye movement desensitization and reprocessing (EMDR)
Emergency Action Plan, whitewater rafting, 186
Emergency response plan (ERP), 58–59
Emergency vehicle running over a pedestrian, case study, 62
EMI, *see* Eye movement integration (EMI) therapy

Emotional effects, of grief and traumatic injuries, 241
Entrapments, 41, 179
　hot tub, 79
　swimming pools, 81–82
　whitewater rafting, 179
Environmental hazards, 179
Equipment (safety)
　boats, 263
　SCUBA, 269–270
　swimming pools, 94
　whitewater rafting, 183–185, 192–194
ERP, *see* Emergency response plan (ERP)
Espinoza v. Arkansas Valley Adventures, LLC, 172
Evidence collection, 63–69
Evidence recovery, 253
Expert (Class V) rapids, 175
Extreme and exploratory (Class VI) rapids, 176
Eye movement desensitization and reprocessing (EMDR), 246
Eye movement integration (EMI) therapy, 246

F

Farrugia, A., 255, 256
Fatal drownings, viiif, vii
Federal Rule of Evidence 702, 69, 70
Fifth Circuit Court of Appeal of Louisiana, 4
Fires, as beach hazard, 148
First aid
　courses, 182–183
　items, 194–195
First responders, 42–43
Flagellate form of *Naegleria fowleri*, 227f, 227–228
Flags, as signs, 52, 54, 54f
Florida
　beaches in, 138
　Breaux v. City of Miami Beach, 197–208
　law on feeding and baiting of sharks, 272
Florida Beach Patrol Chiefs Association, xi
Florida Department of Natural Resources v. Garcia, 202
Flow limits, whitewater rafting, 186
Fluctuating water levels, 177–178
Fluorescent depth marker, 130
Forensic entomologist, 67
Forensic investigation
　body decomposition, stages and description of, 250–252
　determining timeline for drowning event, 256
　diatom test, 254–255
　essentials of, 250–256
　evidence recovery, 253
　initial assessment, 252
　other considerations, 254

　risk–benefit factor, 252
　search for victims, 253
　underwater survey, 252
　victim recovery, 253–254
Fowler, M., 224
Free available chlorine (FAC), 94
Free-living amebae, *see Naegleria fowleri*
Fresh stage, of body decomposition, 250
Fresh *vs.* salt water drowning, 32–33
Freshwater springs, 166–167
Frye standard, 69
Frye v. United States, 69, 70
Ft. Lauderdale new river drownings, case study, 57

G

Gabrielsen, M. Alexander, 102, 112, 113, 116, 119, 122, 133
Gaitlin-Johnson v. City of Miles City, 229
Gallup International poll, 243
Gannon, K., 250, 252
Garber v. Prudential Ins. Co. of Am., 12–13
Gates, swimming pools, 87–88
Gilbertson, D. L., 250, 252
Governmental owners, duties of, 228–232
GPS positioning, 283–284
Grand Canyon National Park, Commerical Outfitter Regulations (COR), 193
Grief/grieving process
　behavioral effects, 242
　contributing factors to, 239
　defined, 238
　emotional effects, 241
　healing/treatment, 244–247
　　CBT, 244–245
　　EMI therapy and EMDR, 246
　　medication, 246–247
　　physical therapy, 247
　　relaxation therapy, 245
　　support groups, 247
　losses and, 238
　mental effects, 241
　physical effects, 240–241
　spiritual effects, 242
　symptoms/symptomatology of, 238, 240
Groins, breakwaters, and jetties, 142–143
Gross negligence, 26
Guides, *see* River guides
Guilt, 242; *see also* Grief/grieving process

H

Harrisburg, 214–215
Hawaii, beaches in, 138
Hazards, xi; *see also* Risk management
　beach, *see* Beach hazards
　diving, 267

Index 295

lakes, 162–163
low-head dams, 276
river, 177–179
Healing/treatment, grief and trauma, 244–247
   CBT, 244–245
   EMI therapy and EMDR, 246
   medication, 246–247
   physical therapy, 247
   relaxation therapy, 245
   support groups, 247
Helical flow, whitewater, 176
HEPCA, *see* Hurgada Environmental Protection and Conservation Association (HEPCA)
High-energy beaches, 138
High seas
   in case law, 217–218
   in DOHSA, 217–218
   in international law, 216–217
High water hazard, 178
Hole, whitewater, *see* Hydraulic/hole, whitewater
Hopper bottom pools, 128
Hotel/mote pools, operation and management of, 134
Hot tub; *see also* Spa; Toddlers drowning
   deaths in, 79
   drownings/accidents, 40, 79
   electrocution, 79
   entrapment injuries, 79
   safety precautions, 79
   signs and safety, 90, 92
   water temperature, 79
Howard v. Equitable Life Assurance Soc. of U.S., 15
Hurgada Environmental Protection and Conservation Association (HEPCA), 274
Hutchinson v. Township of Portage, 11
Hydraulic/hole, whitewater, 177
Hydrodynamic analysis, of drowning events, 281–289
   daily conditions, 286–287
   determining time, 284–285
   initial assessment, 285–286
   introductory conversation, 281–282
   logistics, 282
   model
      configuration and collection of topography, 288
      selection, 287
      validation, 288–289
   trajectory analysis, 284
Hydrologic features, whitewater, 176–177, 176f
   eddy and eddy lines, 176–177
   helical flow, 176
   hydraulic/hole, 177

   laminar flow, 176
   waves, 177

I

Inflatable kayaks, 188
Inflatable rescue boat (IRB), 277
Inground pools, diving accidents, 116–118
Initial apnea, 34
Initial assessment
   forensic investigation, 252
   hydrodynamic conditions, 285–286
Initial assessment, for forensic investigation, 252
Injury(ies); *see also* Accidents
   accidental, vii
   entrapment, *see* Entrapments
   reenactments, 258
   SCI, 81, 102, 108–113
   swimming pools
      case study, 259
      children, 81
      drains and entrapment, 81–82
      secondary factors contributing to, 81
   whitewater rafting, 190–191
Insurance and medical industry, diving accidents and, 134–135
Insurance contract provisions, 15
Intent
   defined, 2
   proof of, 2
Intentional torts, 2
Intermediate (Class III) rapids, 175
International Organization for Standardization (ISO), 266
International Pip Current Symposium in South Korea (2014), 278
International Rafting Federation (IRF), 182
International Scale of River Difficulty, 175
International Swimming Hall of Fame, xi
Interoceptive exposure, 245
IRF, *see* International Rafting Federation (IRF)
ISO, *see* International Organization for Standardization (ISO)

J

Jellyfish, as beach hazard, 156
Jet Ski, 213
Jetties, *see* Groins, breakwaters, and jetties
Jones v. Gillen, 4

K

Kalm, Inc. v. Hawley, 13–14
Keshlear, B., 158
Kübler-Ross, Elizabeth, 242–243
Kübler-Ross model, 242–243

L

Lake Michigan, 158
Lakes, 157–163
    Army Corps and, 158
    for bathing, characteristics of, 158–160
    bottoms, 162
    case study (multiple drownings), 62
    categories of, 157
    lifeguards, 162
    recreational opportunities and value, 157, 158
    responses to hazards, 162–163
    slopes, 160
    warning signs, 160
    water quality, 161
Laminar flow, whitewater, 176
Land management groups, 180
Lauritzen v. Larsen, 215, 219
Law
    practices and principles, vii–ix
    specialties, vii
Layers of protection, 75–76
Leakas v. Columbia Country Club, 13
Learn to Dive, 126, 127, 132, 133, 134
Leonard v. Peoples Camp Corp., 10
Liability
    attractive nuisance doctrine, 14–15
    insurance contract provisions, 15
    intentional torts theory, 2
    negligence theory, 4–14
    other considerations, 14–16
    strict liability theory, 3–4
Liability without fault, *see* Strict liability
Life expectancy, 242
"Lifeguard Effectiveness: A Report of the Working Group," 43
Lifeguards, 43–46
    lakes, 162
    swimming pools, 94–96, 132
Lighting
    beach hazards, 154
    swimming pools, 82, 130
Limitation of Liability Act, *see* Shipowner's Limitation of Liability Act
Lion fish, as beach hazard, 156
LMN, *see* Lower motor neurons (LMN)
Locations of pool, 130
Logos, *see* Warning logos
Lord Campbell's Act (1846), 214
Louganis, Greg, 103–104
Loveland v. Orem City Corp., 4
Low-energy beaches, 138
Lower motor neurons (LMN), 110
Low-head dams, 274–278
    as power sources, 275
    rescue, 277–278
    safety hazards, 276
    types of, 275–276
    warning signs, 276–277, 277f
Low water hazard, 178
Ludes, B., 255, 256
Lysing action, *of Naegleria fowleri*, 225

M

Management, of swimming pools, *see* Operation and management, of swimming pools
Maritime Law Association of the United States (MLAUS), 215
Markings on pool bottoms, 129
McFarland v. Grau, 12
McKinney v. Adams, 199
Medical conditions, whitewater rafting and, 180
Medical industry, *see* Insurance and medical industry, diving accidents and
Medication, 246–247
Mental effects, of grief and traumatic injuries, 241
Metropolitan Dade County v. Zapata, 7, 13
A Million Ways to Die in the West, 1
Mindfulness-based approaches, 245
Minnesota
    case of death caused by *Naegleria fowleri*, 228–232
    as land of lakes, 157–158
MLAUS, *see* Maritime Law Association of the United States (MLAUS)
Model, hydrodynamic conditions
    configuration and collection of topography, 288
    selection, 287
    validation, 288–289
Mottonen, M., 255
Movement-based relaxation methods, 245

N

Naegleria fowleri, 224–232
    attacks, 225, 226f
    chlorination for, 228
    cyst, 227f, 228
    duties of governmental owners and operators, 228–232
    flagellate, 227f, 227–228
    as leading cause of death, 224–228
    lysing action of, 225
    nasal passage and mucosa, 225–227
    overview, 224
    trophozoites, 224–225
    unicellular ameba, 224
Nasal mucosa, *Naegleria fowleri* and, 225–227
National Association of Rescue Divers, xi

Index

National Association of Underwater Instructors (NAUI), 266
National Drowning Prevention Alliance (NDPA), xi
National Oceanographic and Atmospheric Administration (NOAA), xi
National Spa and Pool Institute (NSPI), 133
"The National Summit on Wrongful Convictions: Building a Systematic Approach to Prevent Wrongful Convictions," 64
National Swimming Pool and Spa Institute, 93
National Swimming Pool Foundation (NSPF), xvi
Natural environment, diving accidents in, 123–125
NAUI, *see* National Association of Underwater Instructors (NAUI)
NDBC, *see* U.S. National Data Buoy Center (NDBC)
Near-drowning, vii
Negligence, 4–14
    causation, 5
    comparative, 6
    contributory, 6
    elements, 4
    existence of a duty to exercise reasonable care, 4–5
    lawsuits alleging, 50
Neon nighttime depth marker, 130
Neuropsychological evaluations, 245
Nitrogen narcosis, 269
Non-fatal drownings, xi
Novice (Class II) rapids, 175
NSPF, *see* National Swimming Pool Foundation (NSPF)
NSPI, *see* National Spa and Pool Institute (NSPI)

O

Ocean beaches, *see* Beaches
Ocean forensics, 67–68
Odom v. Lee, 9
Offshore Logistics, Inc. v. Tallentire, 216
Olympic Games, 100, 103
On-scene assessment, 66–67
Operation and management, of swimming pools, 132–134
    by hotel/motel, condo/apartment and public pool, 134
    industry standards, 133
    laws and regulations, 133
    by residential pool owners, 133
    supervision, 132
Outdoor Foundation, 173
Outfalls, as beach hazard, 144–145
Outlet covers, swimming pools, 81–82
Overseas swimming pool safety, 96

Oxygen toxicity, 269

P

PADI, *see* Professional Association of Underwater Instructors (PADI)
PAM (primary amebic meningoencephalitis), 224–225, 228, 230; *see also Naegleria fowleri*
Panic control therapy (PCT), 244
Paraplegia, 110; *see also* Spinal cord injuries (SCI)
Participant limits, whitewater rafting, 185
Passive/silent drowning, 38
People v. Denis, 2
Personal flotation devices (PFD), 77–78
Personal safety equipment, 184–185
PFD, *see* Personal flotation devices (PFD)
Physical effects, of grief and traumatic injuries, 240–241
Physical loss, 239
Physical therapy, 247
Piers, as beach hazard, 141–142
Poleyeff v. Seville Beach Hotel Corporation, 201
Police, 42–43
Polytheism, 243
Pools, *see* Swimming pools
Portuguese man-o-war, 156
Posttraumatic stress disorder (PTSD), vii, xvii, 239
Private pool, 75, 76
Private residential pools, 74
    layers of protection, 76
    operation and management, 133
Professional Association of Underwater Instructors (PADI), 266
Proximate cause, 5, 27–28
Psychological evaluation, 237–247; *see also* Grief/grieving process
Psychotropic medications, 246
PTSD, *see* Posttraumatic stress disorder (PTSD)
Public duty doctrine, 229–230
Public pools, 75; *see also* Swimming pools
    case study of drowning, 61
    coaches and coaching, 78–79
    as competition pools, 78
    layers of protection, 77–78
    operation and management, 134
Pulmonary embolism, 269

Q

Quarries, 165–166
    dangers associated with, 166
    SCUBA, 165

R

Rafting, *see* Whitewater rafting
Rapids; *see also* Whitewater
    Class I (easy), 175
    Class II (novice), 175
    Class III (intermediate), 175
    Class IV (advanced), 175
    Class V (expert), 175
    Class VI (extreme and exploratory), 176
    defined, 174
Record keeping, 59–60
Recovery of victims, 253–254
Recreational headfirst entry accidents, 134
Recreational immunity, 230–232
Recreational lakes, 157, 162; *see also* Lakes
Recreational water illnesses (RWI), 93
Redwood Group, 95
Reenactments, 256–258
    boat injury/drowning, 258
    diving, 257
    SCUBA, 257
Relaxation therapy, 245
Rescue certification, of river guides, 182
Residential pool owners, 133
Restatement of Torts
    Section 430, 231
    Sections 335, 230
    Sections 339, 230–231
Rickwalt v. Richfield Lakes Corp., 7
Rigging, 180
Rip currents, as beach hazard, 150–152
Risk(s)
    identification of, 55–56
    measuring, 56–57
    minimizing and/or eliminating, 58
Risk management, 49–62
    case study, 61–62
    as continuous and ongoing process, 50
    emergency response plan (ERP), 58–59
    in-house and recurrent training, 59
    objective of, 50
    record keeping, 59–60
    response to hazards, 50
    team, 51f
    training, 50
    trespassing, 60–61
    warnings, 51–54
    whitewater rafting, 185–188
        boat order, 187–188
        Emergency Action Plan, 186
        flow limits, 186
        guide selection, 186
        inflatable kayaks, 188
        participant limits, 185
        safety briefing, 187
Risk-taking behavior, 50, 80

Risk–utility analysis, 3
Rivera v. Philadelphia Theological Seminary of St. Charles Borromeo, Inc., 8
River guides, 181–183
    defined, 181
    job of, 181
    selection, 186
    training, 181–183
River hazards
    entrapments, 179
    fluctuating water levels, 177–178
    high water, 178
    low water, 178
    sieves, 179
    strainer, 178
    swimming, 178
RMS Titanic, 216
Ropes, 180
Rules of Evidence (2016), 71
RWI, *see* Recreational water illnesses (RWI)

S

Safety briefing, whitewater rafting, 187
Safety ledges, 131
Safety standards
    swimming pools
        anti-entrapment drains, 81–82
        barrier fences, 85–87
        bathing load, 92–93
        decks, 84
        depth markers, 90
        design and engineering, 88
        diving boards, 85
        equipment, 94
        gates, 87–88
        lifeguard protection, 94–96
        lighting, 82
        outlet covers, 81–82
        overseas safety, 96
        signage, 88–90
        water chemistry, 93–94
        water clarity, 83–84
        waterslides, 88
    whitewater rafting, 180–185
        equipment, 183–185
        river guides, 181–183
Saga Bay Prop. Owners Ass'n v. Askew, 14
Salt *vs.* fresh water drowning, 32–33
Sanchez v. Loffland Brothers Co., 218
SCUBA
    accidents, 265–274
    certification program, 265–266
    decompression sickness, 268–269
    defective equipment, 269–270
    dive training, 266
    drownings, 41–42

Index

nitrogen narcosis, 269
oxygen toxicity, 269
pulmonary embolism, 269
quarries, 165
sea life and, 268
shark diving encounters, 272–274
Sea cliffs, as beach hazard, 157
Sea Grant Foundation, the American Red Cross, xi
Sea life, SCUBA diving and, 268
Secondary drowning, 15–16
Semi-public pools, 74–75
  layers of protection, 76–77
Seville Hotel, 197–198, 200–201, 205; *see also Breaux v. City of Miami Beach*
Shallow water blackout (SWB), 39–40, 270
  factors contributing to, 270
Shallow water diving
  accidents, 125–127
  beach hazards, 152–153
  case study, 258
Sharks/shark attack
  as beach hazard, 155–156
  SCUBA diving encounters, 272–274
Shipowner's Limitation of Liability Act, 214, 216
Sieves, 179
Signs/signage
  colored flags, 52, 54, 54f
  hot tub and spa, 90, 92
  swimming pools, 88–90
Silent drowning, *see* Passive/silent drowning
Simons v. Hedges, 10
Simulating Waves in the Nearshore (SWAN), 287
60 Minutes, x
Skeletal stage, of body decomposition, 251
Smith v. Chicago & E.I.R. Co, 14
Snorkeling, 270, 271f
  drowning, 41–42
Solar depth marker, 130
South Africa
  shark diving in, 272
Spa; *see also* Hot tub
  deaths in, 79
  drownings/accidents, 40, 79
  electrocution, 79
  entrapment injuries, 79
  safety precautions, 79
  signs and safety, 90, 92
  water temperature, 79
Spa drownings, 40
Spady v. Bethlehem Area School Dist., 16
Sperka v. Little Sabine Bay, Inc., 199–200
Spinal cord injuries (SCI), 81, 102, 108–113
  severity of, 110
  statistics, 111–112
  types of, 110
Spiritual effects, of grief and traumatic injuries, 242

Spoon bottom pools, 128
Sport of diving, 100–101, 102
Sport of diving headfirst entry accidents, 135
Sport of swimming headfirst entry accidents, 134–135
Springboard/jumpboard
  competitions, 103
  diving accidents, 118–121
  instructional signs/rules for, 131–132
Springs, 166–167
Squeezes, *see* Barotrauma
Starting blocks, swimming pool
  diving accidents, 121–123
  safety recommendation, 129
Stingrays, as beach hazard, 156
Strainer, 178
Strict liability, 3–4
Suarez v. City of Texas, 7
Submerged Lands Act, 217
Submerged rocks and boulders, 143
Support groups, 247
Surf beaches, 139; *see also* Beaches
Surf zone, 285
SWAN, *see* Simulating Waves in the Nearshore (SWAN)
SWB, *see* Shallow water blackout (SWB)
Swimming hazard, whitewater rafting and, 178
Swimming pools
  case study of drowning, 62
  design and recommended changes, 128–132
    barriers, 130
    color-coded pool edges, 131
    depth marker, 129–130
    lifeline, 131
    lighting, 130
    locations, 130
    marked edges or seats, 131
    markings on bottoms, 129
    recessed steps, ladders, anchors, and receptacles, 131
    safety ledges elimination, 131
    signs for springboard/jumpboard, 131–132
    starting blocks, 129
    warning signs, 130
    water depth awareness, 130
    water slides, 129
  drowning/injuries
    case study, 259
    children, 81
    drains and entrapment, 81–82
    secondary factors contributing to, 81
  number of, 74
  operation and management, 132–134
    by hotel/motel, condo/apartment and public pool, 134
    industry standards, 133

laws and regulations, 133
   by residential pool owners, 133
   supervision, 132
risks, 55
safety
   anti-entrapment drains, 81–82
   barrier fences, 85–87
   bathing load, 92–93
   decks, 84
   depth markers, 90
   design and engineering, 88
   diving boards, 85
   equipment, 94
   gates, 87–88
   lifeguard protection, 94–96
   lighting, 82
   outlet covers, 81–82
   overseas safety, 96
   signage, 88–90
   water chemistry, 93–94
   water clarity, 83–84
   waterslides, 88
types of, 74–75

T

Terminal apnea, 34–35
Tetraplegia, 110; *see also* Spinal cord injuries (SCI)
Third U.S. Circuit Court of Appeals, 213
Timeline determination
   drowning events, 284–285
   during forensic investigation, 256
Toddlers drowning, 40, 80
   reasons, 80
Torts, 26
   breach of duty, 27
   damages, 28
   duty, 27
   elements of, 27–28
   proximate cause, 27–28
TRA, *see* Trinity River Authority (TRA)
Training
   dive, 266
   in-house and recurrent, 59
   risk management, 50
   of river guides, 181–183
      CPR certification, 182
      first aid courses, 182–183
      rescue certification, 182
Trajectory analysis, hydrodynamic condition, 284
Trauma
   cognitive behavioral therapy (CBT), 244–245
   defined, 238
   EMI therapy and EMDR, 246
   medication, 246–247
   physical therapy, 247
   relaxation therapy, 245
   support groups, 247
   symptoms, 238–239
Treatment, grief and trauma, 244–247; *see also* Healing/treatment, grief and trauma
Trespassing, 60–61
Trinity River Authority (TRA), 11
Trinity River Authority v. Williams, 11
TWA Flight 800 crash, 213

U

UMN, *see* Upper motor neurons (UMN)
Under Currents, 265
Underwater survey, for forensic investigation, 252
United States Lifesaving Association (USLA), 43, 44, 47, 139
United States Coast Guard, xi
United States v. Louisiana, 217
Upper motor neurons (UMN), 108
USA Diving, 135
   Dive Safe Manual, 127
   lesson program, 126, 128
   teaching program, 126
   website, 101
U.S. Coast Guard
   on boating accidents, 262
   on boat safety equipment, 263
   on BUI dangers, 264
   on captain's responsibility, 263
U.S. Diving Team, 103
U.S. Forest Service, 180
USLA, *see* United States Lifesaving Association (USLA)
U.S. National Data Buoy Center (NDBC), 286
U.S. Olympic Committee, 135

V

Validation, model, hydrodynamic conditions, 288–289
Victims, forensic investigation
   recovery, 253–254
   search, 253
Vinyl bottom pools, 128–129
Virginia Graeme Baker Pool and Spa Safety Act, 41, 49, 74, 81

W

Ward v. City of Millen, 9
Warning logos, 51, 52f
Warnings, 51–54
   flag, 52, 54, 54f
   reasons why fail, 51–52
   types of, 51

Index

Warning signs, 51–52, 53f
    colored flags, 52, 54, 54f
    lakes, 160
    low-head dams, 276–277, 277f
    swimming pools, 88–90, 130
Water; *see also* Hydrodynamic analysis, of drowning events
    chemistry, 93–94
    disinfection, 94
Water clarity, swimming pools, 83–84
Water pollution, as beach hazard, 154–155
Waterslides, 88
Water temperature
    hot tub and spa, 79
Wave-dominated beaches, 138
Waves, whitewater, 177
Webb v. Hunter, 11–12
White v. Georgia Power Co., 11
Whitewater
    categorization, 175–176
    concept of, 174
    hydrologic features, 176–177, 176f
        eddy and eddy lines, 176–177
        helical flow, 176
        hydraulic/hole, 177
        laminar flow, 176
        waves, 177
Whitewater rafting
    accidents, 189–191
        death causes, 189–190
        fatalities, 189
        injuries, 190–191
    demographics, 173
    estimated outings, 172
    hazards, 174
        cold water, 179
        environmental, 179
        fluctuating water levels, 177–178
        high or low water, 178
        rigging and ropes, 180
        sieves and entrapments, 179
        strainer, 178
        swimming, 178
    overview, 170–172
    participation, 172–173
    risk management, 185–188
        boat order, 187–188
        Emergency Action Plan, 186
        flow limits, 186
        guide selection, 186
        inflatable kayaks, 188
        participant limits, 185
        safety briefing, 187
    safety standards, 180–185
        equipment, 183–185
        river guides, 181–183
Wiley, Robert, 3
Wiley v. Sanders, 3
Work-related drowning, 41
World Congress on Drowning in 2002, 28
World Health Organization, 28
World Recreational SCUBA Training Council (WRSTC), 266
Written warnings, 51; *see also* Warning signs
WRSTC, *see* World Recreational SCUBA Training Council (WRSTC)

Y

Yamaha v. Calhoun, 213
YMCA, 10, 28, 46

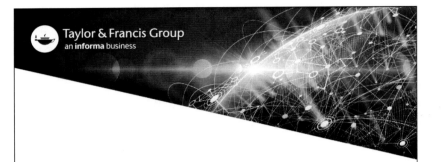